ALGEBRAIC THEORIES

LEONARD E. DICKSON

DOVER PUBLICATIONS, INC.
MINEOLA, NEW YORK

DOVER PHOENIX EDITIONS

Copyright

Copyright © 1926 by Benj. H. Sanborn & Co.
All rights reserved.

Bibliographical Note

This Dover edition, first published in 1959 and reprinted in 2004, is an unabridged and unaltered republication of *Modern Algebraic Theories,* originally published by Sanborn & Co., Chicago, in 1926.

Library of Congress Cataloging-in-Publication Data

Dickson, Leonard E. (Leonard Eugene), 1874-
 [Modern algebraic theories]
 Algebraic theories / Leonard E. Dickson.
  p. cm.—(Dover phoenix editions)
 "This Dover edition, first published in 1959 and reprinted in 2004, is an unabridged and unaltered republication of Modern algebraic theories, originally published by Sanborn & Co., Chicago, in 1926"—T.p. verso.
 Includes index.
 ISBN 0-486-49573-6
 1. Algebra. I. Title. II. Series.

QA155.D5 2004
512—dc22

2004047766

Manufactured in the United States of America
Dover Publications, Inc., 31 East 2nd Street, Mineola, N.Y. 11501

PREFACE

The rapidly increasing number of students beginning graduate work are handicapped by the lack of books in English which provide readable introductions to important parts of mathematics. Nor is the difficulty met adequately by the slow method of lecture courses.

Purpose of this book This book is based on the author's lectures of recent years. It presupposes calculus and elementary theory of algebraic equations. Its aim is to provide a simple introduction to the essentials of each of the branches of modern algebra, with the exception of the advanced part treated in the author's *Algebras and Their Arithmetics*. The book develops the theories which center around matrices, invariants, and groups, which are among the most important concepts in mathematics.

It is a text for several courses The book provides adequate introductory courses in (i) higher algebra, (ii) the Galois theory of algebraic equations, (iii) finite linear groups, including Klein's "icosahedron" and theory of equations of the fifth degree, and (iv) algebraic invariants.

Higher algebra The subject known in America as higher algebra is treated fully in Chapters III–VI; it includes matrices, linear transformations, elementary divisors and invariant factors, and quadratic, bilinear, and Hermitian forms, whether taken singly or in pairs. While the results are classic, the presentation is new and particularly elementary. Due attention is given to questions of rationality, which are too often ignored. The unified treatment of Hermitian and quadratic forms requires but little more space than would be needed for quadratic forms alone. Elementary divisors and invariant factors are introduced in Chapter V in a simple, natural way in connection with the classic and a new rational canonical form of linear transformations; this treatment is not only more elementary than the usual one, but develops these topics in close connection with their most frequent applications. It is then a simple matter to deduce in Chapter VI the theory of the equivalence of pairs of bilinear, quadratic, or Hermitian forms. We thereby avoid the extraneous topic of matrices whose elements are polynomials in a variable and the "elementary transformations" of them.

Algebraic equations Is every equation solvable by radicals? This question is one of absorbing interest in the history of mathematics. It was finally answered in the negative by means of groups of substitutions or permutations of letters. The usual presentation of group theory makes the subject quite abstruse. This impression is avoided in the exposition in Chapters VII–XI. Substitutions are introduced in a very deliberate and natural way in connection with the solution of cubic and quartic equations. The reader will therefore appreciate from the start some of the reasons why substitutions and groups are employed. Fortunately we are able to alternate theory and application in the further exposition of groups. The theory gives very simple answers to the following questions: Can every angle be trisected with ruler and compasses? What regular polygons can be constructed by elementary geometry?

Icosahedron, linear groups Klein's book on the icosahedron and equations of the fifth degree is a classic, but causes real difficulties to beginners on account of the inclusion of ideas from many branches of higher mathematics. Chapter XIII gives a simple exposition of the essentials of this interesting theory, which is a prerequisite to the subjects of elliptic modular functions and automorphic functions. The preliminary Chapter XII discusses the removal of several terms from any equation by means of a rational (Tschirnhaus) transformation, and the reduction of the general equation of the fifth degree to Brioschi's normal form, which is well adapted to solution by elliptic functions. The final Chapter XIV is a sequel to Klein's theory. It establishes remarkable results on the representation of a given group as a

PREFACE   v

linear group, and gives an introduction to Frobenius's theory of group characters. The latter is an effective tool for finite groups and has led to important new results.

Algebraic invariants Chapters I and II provide an easy introduction to the important subject of invariants. Hessians and Jacobians are shown to be covariants and applied to the determination of canonical forms of binary cubic and quartic forms, as well as to the solution of cubic and quartic equations. Every seminvariant is proved to be the leading coefficient of one and only one covariant. It is shown that all covariants of any system of binary forms are expressible in terms of a finite number of the covariants. Such a fundamental system of covariants is actually found for one form of each of the orders 1, 2, 3, 4. Valuable supplementary work on invariants is provided by Chapter XIII, which presupposes the concept of groups of substitutions explained in the elementary Chapter VII.

There are numerous sets of simple problems, and a few historical notes. On pages 38, 133, 176, 203, and 249 there are lists of topics for further reading, with references to writings in English. These topics are suitable for assignment to students for full reports at the end of the particular course.

L. E. DICKSON

University of Chicago
March 16, 1926

CONTENTS

CHAPTER	PAGE

I. Introduction to Algebraic Invariants.......................... 1
Linear transformations. Hessians. Invariants and covariants. Jacobians. Discriminants. Canonical forms of binary cubic and quartic forms. Solution of cubic and quartic equations. Homogeneity. Weights. Seminvariants. Fundamental system of covariants of a binary p-ic for $p < 5$.

II. Further Theory of Covariants of Binary Forms............... 24
Annihilators. Commutators. Existence of a covariant with a given seminvariant leader. Hilbert's theorem. Finiteness of a fundamental system of covariants. Finiteness of syzygies. Canonical form of a binary form of odd order. List of further topics.

III. Matrices, Bilinear Forms, Linear Equations................. 39
Linear forms. Matrices. Linear transformations. Inverse and adjoint matrices. Associative and distributive laws. Characteristic equation. Rank of a matrix. Bilinear forms. Fields. Linear independence. Equivalence of matrices or bilinear forms. Linear equations.

IV. Quadratic and Hermitian Forms, Symmetric and Hermitian Bilinear Forms.. 64
Symmetric bilinear forms and quadratic forms. Hermitian forms and Hermitian bilinear forms. Rational reduction of quadratic and Hermitian forms. Canonical forms. Reduction of a real quadratic form by real orthogonal transformation. Rank of a symmetric or Hermitian matrix. Kronecker's method of reduction. Number of positive coefficients in the canonical form.

V. Theory of Linear Transformations, Invariant Factors and Elementary Divisors....................................... 89
Rational canonical form of a linear transformation. Invariant factors of a canonical transformation. Rotations and orthogonal transformations. Canonical form determined by invariant factors. Similar transformations. Classic canonical form. Elementary divisors.

CHAPTER	PAGE
VI. **Pairs of Bilinear, Quadratic, and Hermitian Forms**...........	112

Equivalence of two pairs of matrices. Canonical forms of a pair of bilinear forms. Pencils of bilinear forms. The nth roots of a matrix. Equivalence of pairs of quadratic or Hermitian forms, or symmetric or Hermitian bilinear forms. Pairs involving alternate forms. Existence of a pair of quadratic or Hermitian forms with any preassigned invariant factors. Weierstrass's canonical pair of quadratic forms. List of further applications of matrices to forms.

VII. **First Principles of Groups of Substitutions**................... 135

Cubic and quartic equations. Their discriminants. Substitutions. Groups. Group leaving a function invariant. Alternating group.

VIII. **Fields, Reducible and Irreducible Functions**.................. 150

Greatest common divisor. Gauss's lemma. Irreducibility of $x^p - A$ and of the cyclotomic equation.

IX. **Group of an Equation for a Given Field**...................... 159

Galois resolvents. Properties A and B of the group of an equation. Transitive and regular groups. Rational functions belonging to a group. Effect on the group by an adjunction to the field. Group of the general equation. Outline of further results.

X. **Equations Solvable by Radicals**............................. 178

History. Resolvent equations and their groups. Invariant subgroup. Transforms of a substitution. Simple and quotient groups. Series and factors of composition. Solvable groups. Equations with a cyclic group. Cyclotomic equations. Theorems of Jordan and Galois. Criterion for solvability by radicals. General equation of degree $n>4$ not solvable by radicals. Solvable quintics.

XI. **Constructions with Ruler and Compasses**..................... 204

Analytic criterion for constructibility. Trisection of an angle. Regular polygons.

XII. **Reduction of Equations to Normal Forms**..................... 210

Tschirnhaus transformations. Principal equations. The Bring-Jerrard normal form. Brioschi's normal form of quintic equations.

CONTENTS

CHAPTER   PAGE

XIII. **Groups of the Regular Solids; Quintic Equations**............ 220
Linear fractional transformation corresponding to a rotation. Tetrahedral, octahedral, and icosahedral groups; their invariants and form problems. Principal quintic resolvent of the icosahedral equation; its identification with any principal quintic. General quintic. Transformation of Brioschi's resolvent into the principal resolvent. Galois group of the icosahedral equation. Further results stated.

XIV. **Representations of a Finite Group as a Linear Group; Group Characters**... 251
Reducible linear groups. Representations. Irreducible and reducible group matrices. Regular group matrix. Group characters. Applications to group matrices. Alternating group on five letters. Computation of group characters.

SUBJECT INDEX... 271

AUTHOR INDEX... 275

MODERN ALGEBRAIC THEORIES

Chapter I

INTRODUCTION TO ALGEBRAIC INVARIANTS

Invariants and covariants play an important rôle in the various parts of modern algebra as well as in geometry. The elementary theory presented in this chapter will meet the ordinary needs in other parts of mathematics. It covers rather fully the subject of invariants and covariants of a homogeneous polynomial in x and y of degree < 5, with application to the solution of cubic and quartic equations. When the degree exceeds 4, most of the covariants are too long to be of real use, unless their symbolic representation is employed.

1. Linear transformations. When two pairs of variables x, y and ξ, η are connected by relations of the form

$$x = a\xi + b\eta, \quad y = c\xi + d\eta, \quad D = \begin{vmatrix} a & b \\ c & d \end{vmatrix} \neq 0,$$

these relations define a linear transformation T of determinant D. Consider another linear transformation U defined by

$$\xi = eX + fY, \quad \eta = gX + hY, \quad \Delta = \begin{vmatrix} e & f \\ g & h \end{vmatrix} \neq 0.$$

The equations obtained by eliminating ξ and η are of the form

$$x = kX + lY, \quad y = mX + nY,$$

and define a linear transformation P which is called the product of T and U, taken in that order, and is denoted by TU. The values of the coefficients are

$$k = ae + bg, \quad l = af + bh, \quad m = ce + dg, \quad n = cf + dh.$$

The determinant of P is found to be equal to $D\Delta$, and hence is not zero. If we solve the equations of T and in the result replace x by X and y by Y, we get

$$\xi = D^{-1}dX - D^{-1}bY, \quad \eta = -D^{-1}cX + D^{-1}aY.$$

These equations define a transformation called the *inverse* of T and denoted by T^{-1}. Since the variables of T^{-1} are the same as those of U, the product TT^{-1} is found by eliminating ξ and η and hence is $x = X$, $y = Y$. The latter is called the *identity transformation*. It is readily verified that also $T^{-1}T$ is the identity transformation.

We shall next prove the associative law

$$TU \cdot V = T \cdot UV,$$

which allows us to write TUV for either product. Let

$$X = pu + qv, \quad Y = ru + sv$$

be the equations of V. The product $TU \cdot V$ is found by eliminating first ξ, η and then X, Y between the equations of T, U, V, while the product $T \cdot UV$ is obtained by eliminating first X, Y and then ξ, η between those equations. In each case we must evidently obtain the same equations expressing x and y as linear functions of u and v.

2. Forms and their classification. A polynomial like

$$ax^3 + bx^2y + cxyz + dxz^2,$$

every term of which is of the same total order (here 3) in the variables x, y, z, is called *homogeneous* in x, y, z. A homogeneous polynomial is called a *form*. According as the number of variables is 1, 2, 3, 4, ..., or q, the form is called *unary, binary, ternary, quaternary,* ..., or *q-ary,* respectively. According as the order of the form is 1, 2, 3, 4, ..., p, it is called *linear, quadratic, cubic, quartic,* ..., *p-ic,* respectively.

For example, the polynomial displayed above is a ternary cubic form, while $ax^2 + bxy + cy^2$ is a binary quadratic form.

3. Hessians. The *Hessian* (named after Otto Hesse) of a function $f(x, y)$ of two variables is the determinant

(1) $$\begin{vmatrix} \dfrac{\partial^2 f}{\partial x^2} & \dfrac{\partial^2 f}{\partial x \partial y} \\ \dfrac{\partial^2 f}{\partial y \partial x} & \dfrac{\partial^2 f}{\partial y^2} \end{vmatrix},$$

whose elements are second partial derivatives. For example, the Hessian of $f = ax^2 + 2bxy + cy^2$ is $4(ac - b^2)$.

The Hessian h of $f(x_1, \ldots, x_q)$ is the determinant of order q in which the elements of the ith row are

(2) $$\frac{\partial^2 f}{\partial x_i\, \partial x_1}, \quad \frac{\partial^2 f}{\partial x_i\, \partial x_2}, \quad \ldots, \quad \frac{\partial^2 f}{\partial x_i\, \partial x_q}.$$

Let f become $F(y_1, \ldots, y_q)$ under the linear transformation

(3) $$x_i = c_{i1} y_1 + \cdots + c_{iq} y_q \qquad (i = 1, \ldots, q),$$

whose determinant is

$$\Delta = \begin{vmatrix} c_{11} & \cdots & c_{1q} \\ \cdots & \cdots & \cdots \\ c_{q1} & \cdots & c_{qq} \end{vmatrix}.$$

The product $h\Delta$ is a determinant of order q whose element in the ith row and jth column is the sum of the products of the elements (2) of the ith row of h by the corresponding elements $c_{1j}, c_{2j}, \ldots, c_{qj}$ of the jth column of Δ and hence is equal to the partial derivative with respect to x_i of

(4)
$$\frac{\partial f}{\partial x_1} c_{1j} + \frac{\partial f}{\partial x_2} c_{2j} + \cdots + \frac{\partial f}{\partial x_q} c_{qj}$$
$$= \frac{\partial f}{\partial x_1} \frac{\partial x_1}{\partial y_j} + \cdots + \frac{\partial f}{\partial x_q} \frac{\partial x_q}{\partial y_j} = \frac{\partial F}{\partial y_j}.$$

Let Δ' denote the determinant obtained from Δ by interchanging its rows and columns, whence $\Delta' = \Delta$. In the product $\Delta' \cdot h \Delta$, the element in the rth row and jth column is therefore

$$c_{1r} \frac{\partial}{\partial x_1} \frac{\partial F}{\partial y_j} + \cdots + c_{qr} \frac{\partial}{\partial x_q} \frac{\partial F}{\partial y_j} = \frac{\partial}{\partial y_r} \frac{\partial F}{\partial y_j},$$

since c_{ir} is the partial derivative of x_i in (3) with respect to y_r. Hence

$$\Delta^2 h = \left| \frac{\partial^2 F}{\partial y_r \partial y_j} \right|_{r, j=1, \ldots, q} = \text{Hessian of } F.$$

THEOREM 1. *If F is the function obtained from f by applying any linear transformation of determinant Δ, the Hessian of F is equal to the product of the Hessian of f by Δ^2.*

4. Definition of invariants and covariants. Let a transformation T of determinant Δ, expressing x and y linearly in terms of ξ and η, replace $f = ax^2 + 2bxy + cy^2$ by $A\xi^2 + 2B\xi\eta + C\eta^2$. We saw in §3 that the Hessian of f is the product of the *discriminant* $ac - b^2$ of f by 4. By Theorem 1,

$$AC - B^2 = \Delta^2(ac - b^2).$$

We therefore call the discriminant $ac - b^2$ an *invariant* of *index* 2 of f. We shall generalize this definition.

§ 4]   DEFINITION OF COVARIANTS   5

Consider the general binary form of order p,

$$f = a_0 x^p + a_1 x^{p-1} y + \cdots + a_p y^p.$$

Let a transformation T replace f by

$$F = A_0 \xi^p + A_1 \xi^{p-1} \eta + \cdots + A_p \eta^p.$$

A polynomial $I(a_0, \ldots, a_p)$ is called an *invariant* of *index* l of f if, for every transformation T of determinant $\Delta \neq 0$,

$$I(A_0, \ldots, A_p) = \Delta^l I(a_0, \ldots, a_p),$$

identically in $a_0, \ldots, a_p$, after the A's have been replaced by their values in terms of the a's.

Covariants K are defined similarly. If, for every transformation T of determinant $\Delta \neq 0$, a polynomial K in the coefficients and variables of f has the property that

$$K(A_0, \ldots, A_p; \xi, \eta) \equiv \Delta^l K(a_0, \ldots, a_p; x, y),$$

identically in $a_0, \ldots, a_p, \xi, \eta$, after the A's have been replaced by their values in terms of the a's, and after x and y have been replaced by their values in terms of ξ and η from T, then K is called a *covariant* of *index* l of f.

By Theorem 1, the Hessian of f is a covariant of index 2 of f. Note that f itself is a covariant of index zero of f since $F = \Delta^0 f$.

By a covariant of index l of a system of forms $f_1, \ldots, f_k$ is meant a function of their coefficients and variables $x_1, \ldots, x_q$ whose product by Δ^l is equal to the same function of the corresponding coefficients and variables $y_1, \ldots, y_q$ in the forms $F_1, \ldots, F_k$ derived from $f_1, \ldots, f_k$ by applying the general linear transformation (3) whose determinant Δ is not zero. An example is given by Theorem 2 below.

A covariant which does not involve any of the variables is an invariant.

5. Jacobians. The *functional determinant* or *Jacobian* (named after C. G. J. Jacobi) of q functions $f_1(x_1, \ldots, x_q), \ldots, f_q(x_1, \ldots, x_q)$ with respect to the variables $x_1, \ldots, x_q$ is defined to be the determinant

$$
(5) \qquad \begin{vmatrix} \dfrac{\partial f_1}{\partial x_1} & \dfrac{\partial f_1}{\partial x_2} & \cdots & \dfrac{\partial f_1}{\partial x_q} \\ \cdots\cdots\cdots\cdots\cdots\cdots\cdots \\ \dfrac{\partial f_q}{\partial x_1} & \dfrac{\partial f_q}{\partial x_2} & \cdots & \dfrac{\partial f_q}{\partial x_q} \end{vmatrix}
$$

THEOREM 2. *The Jacobian of $f_1, \ldots, f_q$ is a covariant of index unity of the system of forms $f_1, \ldots, f_q$.*

Under a transformation (3) of determinant Δ, let f_i become $F_i(y_1, \ldots, y_q)$. Then (4) holds if we put the subscript i on each f and F. Hence the Jacobian of $F_1, \ldots, F_q$ with respect to the variables $y_1, \ldots, y_q$ is equal to

$$
\begin{vmatrix} \dfrac{\partial f_1}{\partial x_1} c_{11} + \cdots + \dfrac{\partial f_1}{\partial x_q} c_{q1} \cdots \dfrac{\partial f_1}{\partial x_1} c_{1q} + \cdots + \dfrac{\partial f_1}{\partial x_q} c_{qq} \\ \cdots\cdots\cdots\cdots\cdots\cdots\cdots\cdots\cdots\cdots\cdots\cdots\cdots\cdots \\ \dfrac{\partial f_q}{\partial x_1} c_{11} + \cdots + \dfrac{\partial f_q}{\partial x_q} c_{q1} \cdots \dfrac{\partial f_q}{\partial x_1} c_{1q} + \cdots + \dfrac{\partial f_q}{\partial x_q} c_{qq} \end{vmatrix}.
$$

This determinant is the product of the determinant (5) by the determinant $\Delta = |c_{ij}|$.

EXERCISES

1. If $f_1, \ldots, f_q$ are dependent functions, their Jacobian is identically zero. [Use $f_q = g(f_1, \ldots, f_{q-1})$.]

2. Two functions f and g of x and y are dependent if their Jacobian J is identically zero.

[Unless g is a constant function of f, we may take $\partial g/\partial x$ not identically zero and hence employ g and y as new variables in place of x and y. Then

§ 6]   DISCRIMINANT OF CUBIC   7

$f = F(g, y)$, and J reduces to $-\dfrac{\partial g}{\partial x}\dfrac{\partial F}{\partial y}$.]

3. If $f(x, y)$ is a binary form of order p, then (Euler)
$$x\frac{\partial f}{\partial x} + y\frac{\partial f}{\partial y} \equiv pf.$$

4. The Hessian of $(ax + by)^p$ is identically zero.
[It is sufficient to prove this for x^p. Why?]

5. Conversely, if the Hessian of a binary form $f(x, y)$ of order p is identically zero, f is the pth power of a linear function.
[The Hessian of f is the Jacobian of $\partial f/\partial x$, $\partial f/\partial y$. The latter are dependent by Ex. 2. Thus
$$b\frac{\partial f}{\partial x} - a\frac{\partial f}{\partial y} \equiv 0,$$
where a and b are constants. Solve this with Euler's relation in Ex. 3, find the derivatives of $\log f$, and integrate.]

6. If the Jacobian of a quadratic form f and a linear form l in x and y is identically zero, f is the product of l^2 by a constant.

7. If a form f in q variables is expressible as a form in fewer than q variables, the Hessian of f is identically zero.[1]

8. The Jacobian of any binary form f and its Hessian h is a covariant of index 3 of f.

9. $f = ax^2 + 2bxy + cy^2$ and $g = px^2 + 2qxy + ry^2$ have the invariant $s = ar - 2bq + cp$ of index 2.
[The discriminant of $f + tg$ is an invariant for every constant t.]

10. A binary quadratic form f and a linear form l have a linear covariant J, their Jacobian. Show that the Jacobian of J and l is an invariant of index 2 of the system f and l.

11. For f, l, J in Ex. 10, prove that the points represented by $f = 0$ are separated harmonically by the points represented by $l = 0$ and $J = 0$.

6. The discriminant of a binary cubic form. Write

(6) $f(x, y) = ax^3 + 3bx^2 y + 3cxy^2 + dy^3.$

[1] The converse is true when q is less than 5, but not when q exceeds 4 (as believed erroneously by Hesse). See Gordan and Nöther, Math. Annalen, 10, 1876, 564.

Its Hessian is $36h$, where

(7) $$h = rx^2 + 2sxy + ty^2, \quad r = ac - b^2, \quad 2s = ad - bc,$$
$$t = bd - c^2.$$

Under any linear transformation of determinant Δ, let f become

(8) $$F = A\xi^3 + 3B\xi^2\eta + 3C\xi\eta^2 + D\eta^3.$$

Denote the Hessian of F by $36H$. Since h is a covariant of index 2 of f,

(9) $$H = R\xi^2 + 2S\xi\eta + T\eta^2 = \Delta^2 h,$$
$$R = AC - B^2,\ 2S = AD - BC,\ T = BD - C^2.$$

Hence $\Delta^2 r$, $2\Delta^2 s$, $\Delta^2 t$ are the coefficients of a binary quadratic form which our transformation replaces by one having the coefficients $R, 2S, T$. Since the discriminant of a binary quadratic form is an invariant of index 2, we have

$$RT - S^2 = \Delta^2\{\Delta^2 r \cdot \Delta^2 t - (\Delta^2 s)^2\} = \Delta^6 (rt - s^2).$$

Hence[1] $rt - s^2$ is an invariant of index 6 of f. We shall call

(10) $$-4(rt - s^2) = (ad - bc)^2 - 4(ac - b^2)(bd - c^2)$$

the *discriminant* of the binary cubic form (6).

To justify this definition, we shall prove

THEOREM 3. *The invariant* (10) *is zero if and only if* $f(x/y, 1)$ $= 0$ *has a multiple root, i.e., if* $f(x, y)$ *is divisible by the square of a linear function of* x *and* y.

For, if the latter be the case, we can transform f into a form (8) having the factor ξ^2; then $C = D = 0$ and the function (10) written in capital letters is zero, so that the invariant (10) itself is zero.

[1] This result and those of Exs. 8 and 10 of §5 illustrate the theorem that any invariant (or covariant) of one or more covariants of a system of forms is an invariant (or covariant) of the forms.

Conversely, let (10) be zero, so that the discriminant of (7) is zero. Then h is the square of a linear function ξ. Apply a linear transformation which expresses x and y in terms of ξ and any function η independent of it. We now have (9) with $h = \xi^2$. Hence $S = 0$, $T = 0$. If $D = 0$, then $C = 0$ (by $T = 0$) and (8) has the factor ξ^2. If $D \neq 0$, we have

$$B = \frac{C^2}{D}, \quad A = \frac{C^3}{D^2}, \quad F = D(\eta + \frac{C}{D}\xi)^3.$$

In either case, $f = F$ is divisible by a square of a linear function.

7. Canonical form of a binary cubic form; solution of cubic equations.

THEOREM 4. *Any binary cubic form can be transformed into $X^3 + Y^3$ if its discriminant is not zero, and into $X^2 Y$ or X^3 if its discriminant is zero.*

Since the last statement has been proved, let the discriminant (10) of the form (6) be not zero. Then, if we discard the factor 36 from all Hessians, the Hessian (7) is the product of two linearly independent factors ξ, η. Hence f can be transformed into a form F of type (8) whose Hessian (9) reduces to $2S\xi\eta$, whence $R = 0$, $T = 0$, $S \neq 0$. If $C = 0$, then $B = 0$ (by $R = 0$) and $F = A\xi^3 + D\eta^3$, where $AD \neq 0$ (by $S \neq 0$). Taking

$$\xi = A^{-\frac{1}{3}} X, \quad \eta = D^{-\frac{1}{3}} Y,$$

we get $F = X^3 + Y^3$, as desired. The remaining case $C \neq 0$ is readily excluded; for, then $B \neq 0$ (by $T = 0$) and

$$A = \frac{B^2}{C}, \quad D = \frac{C^2}{B}, \quad AD = BC, \quad S = 0.$$

Hence to solve a cubic equation $c(w) = 0$ without a multiple root, reduce the cubic form $f(x, y) \equiv y^3 c(x/y)$ to the form

$A\xi^3 + D\eta^3$ by employing the factors ξ and η of the Hessian of f.

EXERCISES

1. The Hessian of $f = x^3 + 6x^2y + 12xy^2 + dy^3$ is $36(d-8)(xy+2y^2)$. When its factors $\xi = x + 2y$ and $\eta = y$ are taken as new variables, f becomes $\xi^3 + (d-8)\eta^3$.
2. For $d = 9$ in Ex. 1, $x/y = -3$, $-2-\omega$, or $-2-\omega^2$, where ω is an imaginary cube root of unity.
3. What are the roots when $d = 7$?
4. The discriminant of $x^3 - 27xy^2 + 54y^3$ is zero.

8. Homogeneity of covariants of a system of forms.

THEOREM 5. *A covariant which is not homogeneous in the variables is a sum of covariants each homogeneous in the variables.*

Let the system of forms have the coefficients $a, b, \ldots$, arranged in any order. If K is a covariant of the forms,

$$K(A, B, \ldots; \xi, \eta, \ldots) = \Delta^l K(a, b, \ldots; x, y, \ldots).$$

When $x, y, \ldots$ are replaced by their linear expressions in $\xi, \eta, \ldots$, the terms of order n_1 on the right (and only terms of order n_1) give rise to terms of order n_1 in $\xi, \eta, \ldots$ on the left. Hence, if K_1 is the sum of all the terms of order n_1 of K, we have

$$K_1(A, B, \ldots; \xi, \eta, \ldots) = \Delta^l K_1(a, b, \ldots; x, y, \ldots),$$

so that K_1 is a covariant of the system of forms. Similarly, the sum K_2 of all the terms of order n_2 of K is a covariant. Hence K is a sum of covariants each homogeneous in the variables.

THEOREM 6. *When a covariant of two or more forms is not homogeneous in the coefficients of each form separately, it is a sum of covariants each homogeneous in the coefficients of each form separately.*

§ 8]   HOMOGENEITY OF COVARIANTS   11

The proof is entirely similar to that used in the following case. Employ the notations of Ex. 9 of §5 and write $d = ac - b^2$. Then $s + 3d$ is an invariant of index 2 of the forms f and g. Let a linear transformation of determinant Δ replace f and g by

$$F = A\xi^2 + 2B\xi\eta + C\eta^2, \qquad G = P\xi^2 + 2Q\xi\eta + R\eta^2,$$

so that A, B, C are linear functions of a, b, c, while P, Q, R are linear functions of p, q, r. By hypothesis,

$$AR - 2BQ + CP + 3(AC - B^2) = \Delta^2(s + 3d).$$

The terms $3d\Delta^2$ of maximum degree 2 in a, b, c on the right arise only from the part $3(AC - B^2)$ on the left. Hence d is itself an invariant of index 2; likewise s is itself an invariant.

In view of Theorems 5 and 6, we may and shall henceforth restrict attention to covariants which are homogeneous in the variables (of constant order n) and homogeneous in the coefficients of each form separately.

THEOREM 7. *Let K be a covariant of order n and index l of the forms $f_1, \ldots, f_r$ of orders $p_1, \ldots, p_r$ in the same variables $x_1, \ldots, x_q$. Let K be of degree d_i in the coefficients of f_i. Then $\sum p_i d_i - n = ql$.*

Let f_i have the coefficients c_{ij} $(j = 1, 2, \ldots)$. Apply the transformation

$$x_1 = my_1, \ldots, x_q = my_q$$

of determinant m^q. Then f_i becomes a form having the coefficients $C_{ij} = m^{p_i} c_{ij}$. Thus

$$K(m^{p_1} c_{1j}, \ldots, m^{p_r} c_{rj}; m^{-1} x_1, \ldots, m^{-1} x_q)$$
$$\equiv (m^q)^l K(c_{1j}, \ldots, c_{rj}; x_1, \ldots, x_q),$$

identically in $m, x_1, \ldots, x_q, c_{1j}, \ldots, c_{rj}$, where each j has its range of values. We may remove the factor m^{-1} and place the factor m^{-n} in front of the left member. Since K is of degree d_i in the c_{ij} ($j = 1, 2, \ldots$),

$$m^{-n}\, m^{p_1 d_1} \cdots m^{p_r d_r} = m^{ql},$$

which proves the theorem.

If, when $r = 1$, we do not assume that K is of constant degree d_1, but consider one term of degree d_1, we see that the proof gives $p_1 d_1 - n = ql$. Hence d_1 is the same for all terms. This proves

THEOREM 8. *A covariant of constant order n of a single form f is of constant degree in the coefficients of f.*

9. Weights of the coefficients of a covariant of a binary form. It is now necessary to employ the subscript notation for the coefficients of the form

(11) $$f = a_0 x^p + p a_1 x^{p-1} y + \cdots + \binom{p}{i} a_i x^{p-i} y^i + \cdots + a_p y^p,$$

the binomial coefficients p, etc., being prefixed to avoid fractions in the later work. The *weight* of a_k is defined to be k and that of a product of factors a_k to be the sum of the weights of the factors.

To make (6) conform with our present notation (11) for $p = 3$, replace a, b, c, d by a_0, a_1, a_2, a_3. Then the Hessian (7) becomes

(12) $$36(rx^2 + 2sxy + ty^2),$$
$$r = a_0 a_2 - a_1^2, \quad 2s = a_0 a_3 - a_1 a_2, \quad t = a_1 a_3 - a_2^2.$$

The terms of r are of constant weight 2, those of s of weight 3, and those of t of weight 4. Since the Hessian is of index 2, this illustrates

WEIGHT OF COVARIANT

THEOREM 9. *In a covariant of order n and index l,*

(13) $$K = \sum_{r=0}^{n} S_r(a) x^{n-r} y^r,$$

of a single form (11), *all terms of the coefficient $S_r(a)$ are of the same weight $l + r$. In particular, the weight of the leader $S_0(a)$ of the covariant is equal to the index l.*

The transformation $x = \xi$, $y = m\eta$ replaces f by a form whose literal coefficients are $A_i = m^i a_i$. By the definition of a covariant K of index l, we have

(14) $$\sum_{r=0}^{n} S_r(A) \xi^{n-r} \eta^r = m^l \sum S_r(a) x^{n-r} y^r.$$

Let $t = c a_0^{e_0} a_1^{e_1} \cdots a_p^{e_p}$ be any term of $S_r(a)$; the weight of t is $w_r = e_1 + 2e_2 + \cdots + p e_p$. The left member of (14) has the term

$$c A_0^{e_0} A_1^{e_1} \cdots A_p^{e_p} \xi^{n-r} \eta^r = t x^{n-r} y^r m^{w_r - r},$$

which must be equal to a term of the right member of (14). Hence $w_r - r = l$. This proves Theorem 9.

Each $S_r(a)$ is *isobaric*, being of constant weight. If we attribute the weight 1 to x and the weight 0 to y, we see that every term of (13) is of total weight $l + n$.

COROLLARY. *A homogeneous polynomial is a covariant of f with respect to every transformation $x = \xi$, $y = m\eta$, if and only if it is isobaric.*

EXERCISES

1. Verify Theorem 9 for the covariant f of f.

2. For an invariant of index l and degree d of one binary form of order p, $pd = 2l$. Here l is also the weight of the invariant.

3. Hence no binary form of odd order has an invariant of odd degree.

10. Seminvariants. A homogeneous, isobaric polynomial in the coefficients of a binary form f is called a *seminvariant* of f if it is an invariant of f with respect to all transformations of the type

$$T_k: \quad x = \xi + k\eta, \quad y = \eta.$$

Let $A_0, A_1, \ldots$ be the literal coefficients of the form obtained from (11) by applying transformation T_k. Although not essential to our discussion, note that

(15) $\quad A_0 = a_0, \quad A_1 = a_1 + ka_0, \quad A_2 = a_2 + 2ka_1 + k^2 a_0.$

The first equation shows that a_0 is a seminvariant of f. Elimination of k between the last two equations gives

$$A_0 A_2 - A_1^2 = a_0 a_2 - a_1^2,$$

whence $a_0 a_2 - a_1^2$ is a seminvariant of f. It is the leader of the Hessian of f for $p = 3$, as noted before Theorem 9. Evidently a_0 is the leader of the covariant f of f. These two cases illustrate

THEOREM 10. *The leader of every covariant K of any binary form f is a seminvariant of f.*

The leader $S_0(a)$ of any covariant (13) is isobaric by Theorem 9 and is homogeneous by Theorem 8. Since K is a covariant of f with respect to every transformation T_k of determinant unity, we have

$$K = \sum S_r(A) \xi^{n-r} \eta^r = \sum S_r(A)(x - ky)^{n-r} y^r.$$

A comparison of the coefficients of x^n with that in (13) gives $S_0(a) = S_0(A)$. Hence $S_0(a)$ is an invariant of f with respect to every transformation T_k. This completes the proof of Theorem 10.

By T_k, $x - ry = \xi - (r - k)\eta$. Hence each root r_i of $f(x/y, 1) = 0$ is diminished by k. Thus the difference of any two roots is unaltered.

Since $A_1 = 0$ in (15) if we take $k = -a_1/a_0$, we see that the resulting transformation T_k replaces f by the reduced form

$$g = a_0 \xi^p + \binom{p}{2} b_2 \xi^{p-2} \eta^2 + \binom{p}{3} b_3 \xi^{p-3} \eta^3 + \cdots,$$

where

$$b_2 = a_2 - a_1^2/a_0, \quad b_3 = a_3 - 3a_1 a_2/a_0 + 2a_1^3/a_0^2.$$

The roots of $g = 0$ are

$$r_i + a_1/a_0 = [(r_i - r_1) + \cdots + (r_i - r_p)]/p$$
$$(i = 1, \ldots, p).$$

Hence each root of $g = 0$ is a linear function of the differences of the roots of $f = 0$ and thus is unaltered by every transformation T_k. The latter is true also of $b_2/a_0, b_3/a_0, \ldots$, which are equal to numerical multiples of the elementary symmetric functions of the roots of $g = 0$. Hence the homogeneous isobaric polynomials

(16) $\begin{cases} S_2 = a_0 b_2 = a_0 a_2 - a_1^2, \\ S_3 = a_0^2 b_3 = a_0^2 a_3 - 3a_0 a_1 a_2 + 2a_1^3, \\ S_4 = a_0^3 b_4 = a_0^3 a_4 - 4a_0^2 a_1 a_3 + 6a_0 a_1^2 a_2 - 3a_1^4, \end{cases}$

etc., to S_p are invariants of f with respect to all transformations T_k, and hence are seminvariants of f.

Since g was derived from f by a linear transformation of determinant unity, every seminvariant S of f has the property

$$S(a_0, \ldots, a_p) = S(a_0, 0, b_2, \ldots, b_p)$$
$$= S(a_0, 0, S_2/a_0, \ldots, S_p/a_0^{p-1}).$$

THEOREM 11. *Every seminvariant of f is the quotient of a polynomial in the seminvariants $a_0, S_2, \ldots, S_p$ by a power of a_0.*

We shall presently apply this theorem to obtain all seminvariants when $p < 5$. Among them occur the leaders of all covariants by Theorem 10. Our next problem will then be to find all covariants K having a given leader S. As an aid to its solution, we now prove

THEOREM 12. *If S is a seminvariant of degree d and weight w of a binary form f of order p, any covariant K of f having the leader S is of known order $n = pd - 2w$, and at most one such K exists.*[1]

By Theorem 9, the weight w of S is the index l of K. By Theorem 7 with $r = 1$, $q = 2$, we have $pd - n = 2l$. Hence the order n of K is uniquely determined by S.

If there exist two distinct covariants $S x^n + \cdots$ of f, they have the same index $l = w$, whence their difference is a covariant of f having the factor y. It is of the form $yq(a_0, \ldots, a_p; x, y)$. Let η be any chosen linear function of x and y and select another such function ξ which is independent of η. In view of the transformation of determinant Δ which expresses x and y linearly in terms of ξ and η, we have

$$\eta q(A_0, \ldots, A_p; \xi, \eta) = \Delta^l yq(a_0, \ldots, a_p; x, y).$$

Hence yq is of order n and has as a factor every η. This contradiction completes the proof of Theorem 12.

COROLLARY. *The weight of an invariant of degree d of a binary form of order p is $\frac{1}{2}pd$.*

11. Theorem 13. *Every binary linear transformation is a product of transformations of the three types*

$T_n:$ $x = \xi + n\eta,$ $y = \eta;$
$S_m:$ $x = \xi,$ $y = m\eta \, (m \neq 0);$
$V:$ $x = -\eta,$ $y = \xi.$

From these we obtain $V^3 = V^{-1}: x = \eta, y = -\xi,$ and

---

[1] That one such K always exists is proved in §16.

§ 11]   GENERATORS OF TRANSFORMATIONS   17

$$R_n = V^{-1} T_{-n} V: \quad x = X, \quad y = Y + nX;$$
$$P_m = V^{-1} S_m V: \quad x = mX, \quad y = Y \quad (m \neq 0).$$

Then $x = a\xi + b\eta,\ y = c\xi + d\eta,\ D = ad - bc \neq 0$, is equal to

$$S_d\, P_{D/d}\, T_{bd/D}\, R_{c/d} \quad (d \neq 0), \qquad S_c\, P_{-b}\, T_{-a/b}\, V \quad (d = 0),$$

Applying the corollary to Theorem 9, and noting the effect of transformation V of determinant unity, we obtain the

COROLLARY. *A polynomial in* $a_0, \ldots, a_p, x, y$ *which is homogeneous in* x *and* y *is a covariant of the p-ic* (11) *if and only if it is isobaric, is covariant with respect to every* T_n, *and is unaltered when* x *is replaced by* y, y *by* $-x$, *and* a_i *by* $(-1)^i a_{p-i}$ *for* $i = 0, 1, \ldots, p$. Hence a seminvariant which is unaltered by the last replacements is an invariant.

12. Fundamental system of covariants of the binary p-ic for p < 5. We shall exhibit a finite number of polynomial covariants $K_1, \ldots, K_n$ of the binary p-ic f such that every polynomial covariant of f is expressible as a polynomial in $K_1, \ldots, K_n$ with numerical coefficients. Then $K_1, \ldots, K_n$ are said to form a *fundamental system* of covariants of f.

We shall first exhibit certain seminvariants of f such that every seminvariant is a polynomial in them. Each of them will be seen to be the leader of a known covariant of f. Our problem is thereby solved, since no seminvariant is the leader of two covariants by Theorem 12.

For $p = 1$, Theorem 11 shows that every seminvariant of $f = a_0 x + a_1 y$ is a polynomial in a_0. Since a_0 is the leader of the covariant f of f, we have

THEOREM 14. *Every polynomial covariant of* $f = a_0 x + a_1 y$ *is a polynomial in* f. *Thus* f *itself constitutes the fundamental system of covariants of* f.

For $p = 2$, Theorem 11 shows that every seminvariant is the quotient of a polynomial $P(a_0, S_2)$ by a_0^e. But S_2 is not divisible by a_0. Hence P is divisible by a_0 only when all of its terms have the explicit factor a_0. Cancelling the factor a_0 from both P and a_0^e, we reduce the problem to a like one having $e - 1$ in place of e, and finally to the case $e = 0$. Hence every seminvariant is a polynomial in a_0 and S_2. Since a_0 is the leader of the covariant f of f, while S_2 is an invariant of f by §4, we have

THEOREM 15. *Every polynomial covariant of the binary quadratic form f is a polynomial in f and its discriminant S_2.*

A different treatment is necessary when $p > 2$ since there are seminvariants which are not polynomials in a_0 and the S_i. Write Σ_i for the terms of S_i free of a_0; by (16)

(17) $\qquad \Sigma_2 = -a_1^2, \quad \Sigma_3 = 2a_1^3, \quad \Sigma_4 = -3a_1^4.$

For $p = 3$, $4\Sigma_2^3 + \Sigma_3^2 \equiv 0$. We find that

(18) $\qquad\qquad 4S_2^3 + S_3^2 \equiv a_0^2 D,$

where, in accord with (10) and (12),

(19) $D = a_0^2 a_3^2 - 6a_0 a_1 a_2 a_3 + 4a_0 a_2^3 + 4a_1^3 a_3 - 3a_1^2 a_2^2$

is the discriminant of the cubic form f. By means of (18) we may eliminate S_3^2 and higher powers of S_3 from any polynomial in a_0, S_2, S_3 and conclude from Theorem 11 that every seminvariant is of the form P/a_0^e, where P is a polynomial in a_0, S_2, S_3, D of degree 0 or 1 in S_3. If $e = 0$, P is a seminvariant. Henceforth let $e \geqq 1$. In case every term of P has the explicit factor a_0, we cancel it from P and a_0^e and hence reduce the problem to a like one having $e - 1$ in place of e. It remains only to treat the case in which not every term of P has the explicit factor a_0. Then the

§ 12] FUNDAMENTAL SYSTEM OF COVARIANTS 19

expression for P in terms of a_0, a_1, a_2, a_3 is not divisible by a_0 (i.e., P does not have the implicit factor a_0). For, if so, then

$$P' \equiv P(0, \Sigma_2, \Sigma_3, d) \equiv 0, \qquad d = 4a_1^3 a_3 - 3a_1^2 a_2^2.$$

Since a_3 occurs in d, but in neither Σ_2 nor Σ_3, we conclude that P' is free of d. The first power of $\Sigma_3 = 2a_1^3$ is not cancelled by a polynomial in $\Sigma_2 = -a_1^2$. Hence P' is free also of Σ_3 and therefore of Σ_2.

Hence every seminvariant of the cubic form f is a polynomial in a_0, S_2, S_3, D. They are connected by the *syzygy* (18), i.e., a relation not serving to express any one of the four as a polynomial in the remaining three.

THEOREM 16. *A fundamental system of polynomial covariants of the binary cubic form f is given by f, its discriminant D, its Hessian $36H$, and the Jacobian $3G$ of f and H. They are connected by the syzygy*

(20) $$4H^3 + G^2 \equiv f^2 D.$$

This follows since a_0, S_2, S_3 are leaders of the covariants f, H, G.

For $p = 4$, we see from (17) that the simplest relation is $3\Sigma_2^2 + \Sigma_4 \equiv 0$. We find that

$$S_4 + 3S_2^2 \equiv a_0^2 I, \qquad I = a_0 a_4 - 4a_1 a_3 + 3a_2^2.$$

We employ this identity to eliminate S_4. Consider polynomials $P(a_0, S_2, S_3, I)$ having a_0 as an implicit, but not explicit, factor. Such a polynomial is given by (18). For $a_0 = 0$, $D = -a_1^2 I = \Sigma_2 I$; we find that

$$S_2 I - D \equiv a_0 J,$$
$$J = a_0 a_2 a_4 - a_0 a_3^2 + 2a_1 a_2 a_3 - a_1^2 a_4 - a_2^3.$$

Elimination of D between this identity and (18) gives

(21) $\qquad a_0^3 J - a_0^2 S_2 I + 4S_2^3 + S_3^2 \equiv 0.$

The latter enables us to eliminate S_3^2 and higher powers of S_3 from P to obtain a polynomial Q in a_0, S_2, S_3, I, J of degree 0 or 1 in S_3. If possible, let Q have a_0 as an implicit, but not explicit, factor. Then

$$Q(0, -a_1^2, 2a_1^3, 3a_2^2 - 4a_1 a_3, J) \equiv 0.$$

In view of the term $-a_1^2 a_4$ of J, Q cannot involve J, and similarly not I. Nor can Q be of degree 1 in S_3 in view of the odd power a_1^3. Hence Q is free also of S_3 and therefore of S_2.

Hence every seminvariant of the quartic f is a polynomial in a_0, S_2, S_3, I, J. They are connected by the syzygy (21).

Denote the Hessian of f by $144H$, and the Jacobian of f and H by $8G$. Then a_0, S_2, and S_3 are the leaders of the covariants f, H and G of f. By the corollary in §11, the seminvariants I and J are invariants of f, since they are unaltered when we replace a_0 by a_4, a_4 by a_0, a_1 by $-a_3$, and a_3 by $-a_1$.

THEOREM 17. *A fundamental system of polynomial covariants of the binary quartic form f is given by f, its invariants I and J, its Hessian $144H$, and the Jacobian $8G$ of f and H. They are connected by the syzygy*

(22) $\qquad f^3 J - f^2 HI + 4H^3 + G^2 \equiv 0.$

13. Canonical form of a binary quartic form f; solution of quartic equations. We shall exclude the uninteresting case in which $f(x/y, 1) = 0$ has a multiple root. Since there are four distinct roots, $f(x, y)$ is a product of four linear forms, no two of which have proportional coefficients. Taking any two of them as new variables ξ and η, we may reduce f by a linear transformation to the form $\xi \eta q$, where

$$q = g\xi^2 + 2h\xi\eta + k\eta^2.$$

Then $gk \neq 0$ since neither ξ nor η is a factor of q. Choose r, s, t so that

$$t^2 = k/g, \quad r + s = g, \quad r - s = h/t.$$

Then

$$\xi\eta = \tfrac{1}{4}t^{-1}[(\xi + t\eta)^2 - (\xi - t\eta)^2],$$
$$q = r(\xi + t\eta)^2 + s(\xi - t\eta)^2.$$

Introducing the new variables

$$X = \sqrt[4]{\frac{r}{4t}}(\xi + t\eta), \quad Y = \sqrt[4]{\frac{-s}{4t}}(\xi - t\eta),$$

we see that $f = \xi\eta q$ takes the canonical form

(23) $$X^4 + Y^4 + 6m X^2 Y^2.$$

The presence of the arbitrary constant m is explained by the existence of a rational *absolute* invariant I^3/J^2, i.e., one of index zero, which therefore has the same value for f as for any form, like (23), derived from f by linear transformation. For, the invariants I and J are of weights (and hence indices) 4 and 6, so that I^3/J^2 is of index zero.

In the preceding proof of the existence of a canonical form (23), we made use of the roots of $f(x/y, 1) = 0$. We shall now show how to obtain the canonical form without assuming that the roots are known. We have proved the existence of a number m and a linear transformation T of determinant $\Delta \neq 0$ which expresses X and Y linearly in terms of x and y, such that (23) becomes the given quartic f under this transformation T. To compute the values of I and J for (23), take $a_0 = 1$, $a_1 = 0$, $a_2 = m$, $a_3 = 0$, $a_4 = 1$;

then I and J have the values $1 + 3m^2$ and $m - m^3$ respectively. Since I and J have the indices 4 and 6, we get

$$f = X^4 + Y^4 + 6m\, X^2\, Y^2,$$
$$I = \Delta^4(1 + 3m^2), \quad J = \Delta^6(m - m^3),$$

where I and J refer to f. The Hessian of (23) is

$$144 \begin{vmatrix} X^2 + mY^2 & 2m\, XY \\ 2mXY & Y^2 + mX^2 \end{vmatrix}.$$

Hence if the Hessian of f is denoted by $144H$, a covariant of index 2, we have

$$H = \Delta^2 \{m(X^4 + Y^4) + (1 - 3m^2)X^2\, Y^2\}.$$

We first compute a root $\Delta^2\, m$ of the cubic equation

(24) $\qquad 4(\Delta^2\, m)^3 - I(\Delta^2\, m) + J = 0.$

Then $\Delta^4 = I - 3(\Delta^2\, m)^2$. Choose either square root as Δ^2. From the value of $\Delta^2\, m$, we get the value of m. Eliminating $X^4 + Y^4$ between the expressions for f and H, we get

$$\Delta^2\, mf - H = \Delta^2(9m^2 - 1)X^2\, Y^2.$$

This gives XY by extracting the square root of a form which is a perfect square. For, if $9m^2 = 1$, f would be the square of $X^2 \pm Y^2$, contrary to the hypothesis that there is no multiple root. From XY and f, we get $X^2 + Y^2$ and then $X \pm Y$ and hence X and Y.

THEOREM 18. *Every binary quartic form $f(x, y)$ not divisible by the square of a linear form can be reduced to a canonical form* (23). *We can find m and the linear functions X and Y of x and y by finding a root r of a cubic equation* (24), *whose coefficients are the invariants I and J of f, and by extracting the square root of $rf - H$, where $144H$ is the Hessian of f.*

We can express (23) as a product of linear factors by solving $z^2 + 1 + 6mz = 0$, where $z = X^2/Y^2$.

THEOREM 19. *To solve any quartic equation $q(w) = 0$ not having a multiple root, reduce the quartic form $f(x, y) = y^4 q(x/y)$ to its canonical form (23) and express it as a product of linear factors by employing square roots of known numbers. To solve the auxiliary cubic equation (24), reduce the corresponding cubic form to one involving only the cubes of the linear factors of the Hessian of that cubic form (§7). Hence we can solve any quartic equation by extracting a cube root and certain square roots of numbers determined by the invariants I and J and the Hessian of the corresponding quartic form f.*

For tables of invariants and covariants, see Faà di Bruno, Theorie der Binären Formen, Leipzig, 1881.

Exercises

1. If $I = 0$, $\Delta \neq 0$, then $3m^2 = -1$. Show that $I = 0$ for the quartic $w^4 - Sw^2 - \frac{2}{3}Tw - \frac{1}{12}S^2 = 0$ (which arises in the theory of inflexion points on a plane cubic curve). Find its quadratic factors.

2. Find the three canonical quartics when $J = 0$. Apply to solve $w^4 + cw^3 \pm cw - 1 = 0$.

Chapter II

FURTHER THEORY OF COVARIANTS OF BINARY FORMS

We shall here prove some fundamental theorems on invariants and covariants which were merely verified in the introductory chapter for binary forms of orders ≤ 4. For example, we shall prove that all covariants are expressible as polynomials in a finite number of covariants. Suggestions for supplementary reading are made at the end of the chapter.

14. Annihilators of covariants. Consider a binary form

(1) $$f = \sum_{i=0}^{p} \binom{p}{i} a_i x^{p-i} y^i,$$

having prefixed binomial coefficients, as in §9. Under the transformation

$$T_n: \quad x = \xi + n\eta, \quad y = \eta,$$

let f become

(2) $$f = \sum_{i=0}^{p} \binom{p}{i} A_i \xi^{p-i} \eta^i,$$

where the A_i are polynomials in $n, a_0, \ldots, a_p$ whose actual expressions will not be needed. Differentiating (2) with respect to n, we get

$$0 = \sum_{i=0}^{p} \binom{p}{i} \left\{ \frac{\partial A_i}{\partial n} \xi^{p-i} \eta^i - A_i (p-i) \xi^{p-i-1} \eta^{i+1} \right\},$$

since $\eta = y$ is free of n, while $\xi = x - n\eta$. The total coefficient of $\xi^{p-j} \eta^j$ is

$$\binom{p}{j} \frac{\partial A_j}{\partial n} - \binom{p}{j-1}(p-j+1)A_{j-1} = 0,$$

§ 14] ANNIHILATORS OF COVARIANTS 25

in which the second term is to be suppressed when $j = 0$. Since

$$\binom{p}{j} = \binom{p}{j-1}\frac{(p-j+1)}{j},$$

we evidently get

(3) $\quad \dfrac{\partial A_0}{\partial n} = 0, \quad \dfrac{\partial A_j}{\partial n} = jA_{j-1} \quad (j = 1, \ldots, p).$

Consider the polynomials

$$k \equiv K(a_0, \ldots, a_p; x, y), \quad K \equiv K(A_0, \ldots, A_p; \xi, \eta).$$

By (3),

(4) $\quad \dfrac{\partial K}{\partial n} = \sum_{j=0}^{p} \dfrac{\partial K}{\partial A_j}\dfrac{\partial A_j}{\partial n} + \dfrac{\partial K}{\partial \xi}\dfrac{\partial \xi}{\partial n} = \sum_{j=1}^{p} jA_{j-1}\dfrac{\partial K}{\partial A_j} - \eta\dfrac{\partial K}{\partial \xi}.$

The polynomial k is said to be a covariant of (1) with respect to every transformation T_n (of determinant unity) if and only if $K = k$ becomes an identity in $n, x, y, a_0, \ldots, a_p$ after ξ is replaced by $x - ny$, η by y, and each A_i is replaced by its expression in terms of $n, a_0, \ldots, a_p$. Since k is free of n, $K = k$ implies that

$$\dfrac{\partial K}{\partial n} \equiv 0.$$

Conversely, the latter implies that K is independent of n and hence is equal to the value k to which it reduces when $n = 0$, since T_n is then the identity transformation $x = \xi$, $y = \eta$, which replaces (1) by a form (2) with the same coefficients $A_i = a_i$. The condition is therefore that the final expression in (4) be zero identically in $n, x, y, a_0, \ldots, a_p$. In particular, it must be zero when $n = 0$, whence

(5) $\quad \Omega k - y\dfrac{\partial k}{\partial x} \equiv 0, \quad \Omega = \sum_{j=1}^{p} ja_{j-1}\dfrac{\partial}{\partial a_j},$

identically in $x, y, a_0, \ldots, a_p$. From this identity we evidently obtain an identity when we replace $x, \ldots, a_p$ by any quantities such as $\xi, \eta, A_0, \ldots, A_p$. Thus the final expression (4) is zero identically in the latter quantities. Hence k is a covariant of (1) with respect to every transformation T_n if and only if (5) holds. We shall then say that k is *annihilated* by the differential operator

(6) $\qquad \Omega - y \dfrac{\partial}{\partial x} \qquad \left(\Omega = a_0 \dfrac{\partial}{\partial a_1} + \cdots + p a_{p-1} \dfrac{\partial}{\partial a_p} \right).$

The form (1) is unaltered if we interchange x and y, and interchange a_i and a_{p-i} for $i = 0, 1, \ldots, p$. Hence k is a covariant of (1) with respect to every transformation

$$R_n: \quad x = \xi, \quad y = \eta + n\xi,$$

if and only if k is annihilated by the operator

(7) $\qquad O - x \dfrac{\partial}{\partial y}$
$\qquad \left(O = p a_1 \dfrac{\partial}{\partial a_0} + \cdots + (p - i) a_{i+1} \dfrac{\partial}{\partial a_i} + \cdots + a_p \dfrac{\partial}{\partial a_{p-1}} \right).$

By the corollary in §9, k is a covariant of (1) with respect to every transformation S_m: $x = \xi$, $y = m\eta$, if and only if k is isobaric. Since $T_{-1} R_1 T_{-1} = V$, viz., $x = -Y$, $y = X$, it follows from §11 that every binary transformation is generated by the T_n, R_n, S_m.

THEOREM 1. *A polynomial $K(a_0, \ldots, a_p; x, y)$ is a covariant of the binary form (1) if and only if it is isobaric and is annihilated*[1] *by the two operators (6) and (7).*

COROLLARY 1. *A polynomial in $a_0, \ldots, a_p$ is an invariant of (1) if and only if it is isobaric and is annihilated by Ω and O.*

---

[1] In computations, it is simpler to employ the single annihilator (6). We apply the corollary in §11 with condition concerning T_n replaced by the equivalent condition that the polynomial be annihilated by (6).

For example, the weight of an invariant I of degree 2 of $a_0 x^2 + 2a_1 xy + a_2 y^2$ is 2 by the corollary at the end of §10. Hence $I = ra_0 a_2 + sa_1^2$. Then $\Omega I = 2(r+s)a_0 a_1 \equiv 0$, $OI = 2(r+s) \cdot a_1 a_2 \equiv 0$. Either condition gives $s = -r$, $I = r(a_0 a_2 - a_1^2)$.

By the definition of seminvariant in §10, we have

COROLLARY 2. *A homogeneous, isobaric polynomial in $a_0, \ldots, a_p$ is a seminvariant of (1) if and only if it is annihilated by Ω.*

15. Commutators. Express each of the annihilators Ω and O of invariants in the following two forms:

$$\Omega = \sum_{j=1}^{p} j a_{j-1} \frac{\partial}{\partial a_j} = \sum_{k=0}^{p-1} (k+1) a_k \frac{\partial}{\partial a_{k+1}},$$

$$O = \sum_{j=1}^{p} (p-j+1) a_j \frac{\partial}{\partial a_{j-1}} = \sum_{k=0}^{p-1} (p-k) a_{k+1} \frac{\partial}{\partial a_k}.$$

Then

$$\Omega O = \sum_{j=1}^{p} j a_{j-1} \left[(p-j+1) \frac{\partial}{\partial a_{j-1}} + \sum_{k=0}^{p-1} (p-k) a_{k+1} \frac{\partial^2}{\partial a_j \partial a_k} \right],$$

$$O\Omega = \sum_{k=0}^{p-1} (p-k) a_{k+1} \left[(k+1) \frac{\partial}{\partial a_{k+1}} + \sum_{j=1}^{p} j a_{j-1} \frac{\partial^2}{\partial a_k \partial a_j} \right].$$

The terms involving second derivatives are identical. Thus $\Omega O - O\Omega$ involves only first derivatives and is called the *commutator* (alternant) of Ω with O. In the first terms of ΩO, write $j = i+1$; in those of $O\Omega$, write $k = i-1$. Hence

$$\Omega O - O\Omega = \sum_{i=0}^{p-1} (i+1) a_i (p-i) \frac{\partial}{\partial a_i} - \sum_{i=1}^{p} (p-i+1) a_i i \frac{\partial}{\partial a_i}$$

$$= \sum_{i=0}^{p} (p - 2i) a_i \frac{\partial}{\partial a_i}.$$

If S is a homogeneous function of $a_0, \ldots, a_p$ of (total) degree d and hence is a sum of terms of type

$$t \equiv c a_0^{e_0} a_1^{e_1} \cdots a_p^{e_p} \qquad (e_0 + e_1 + \cdots + e_p = d),$$

we readily verify Euler's theorem (cf. Ex. 3, §5):

$$\sum_{i=0}^{p} a_i \frac{\partial S}{\partial a_i} \equiv dS.$$

If S is isobaric, it is a sum of terms t in which

$$e_1 + 2e_2 + \cdots + pe_p = w = \text{constant}.$$

Then

$$\sum_{i=0}^{p} ia_i \frac{\partial t}{\partial a_i} \equiv \sum_{i=0}^{p} ie_i\, t \equiv wt, \qquad \sum_{i=0}^{p} ia_i \frac{\partial S}{\partial a_i} \equiv wS.$$

Theorem 2. *If S is both homogeneous (of degree d) and isobaric (of weight w) in $a_0, \ldots, a_p$, then*

(8) $\qquad (\Omega O - O\Omega)S \equiv nS, \qquad n = pd - 2w.$

This result is the case $r = 1$ of

(9) $\qquad (\Omega O^r - O^r \Omega)S \equiv r(n - r + 1)O^{r-1} S.$

To prove the latter by induction on r, assume that it holds when $r = k$, and note that OS is of degree d and weight $w + 1$, so that when (9) is employed with S replaced by OS, n must be replaced by $n - 2$. Hence

$$(\Omega O^{k+1} - O^{k+1} \Omega)S \equiv (\Omega O^k - O^k \Omega)OS + O^k(\Omega O - O\Omega)S$$
$$\equiv k(n - 2 - k + 1)O^{k-1} OS + nO^k S$$
$$\equiv (k + 1)(n - k)O^k S,$$

and (9) holds for $r = k + 1$. The induction is therefore complete.

Similarly, since ΩS is of degree d and weight $w - 1$,

(10) $\quad (O\Omega^r - \Omega^r O)S \equiv r(-n - r + 1)\Omega^{r-1} S.$

16. Existence of a covariant with a given seminvariant leader.

In §12 we verified for $p \leq 4$ that every seminvariant of the binary form f of order p is the leader of a covariant of f. We shall now prove that a like result holds for any p.

LEMMA. *If S is a seminvariant, not identically zero, of degree d and weight w of a binary p-ic, then $pd - 2w \geq 0$.*

Suppose on the contrary that S is a seminvariant for which $n = pd - 2w$ is negative. Since $\Omega S = 0$, (9) gives

(11) $\quad \Omega O^r S = r(n - r + 1)O^{r-1} S \quad (r = 1, 2, \ldots.)$

and no one of the coefficients on the right is zero. We recall that the weight of OP exceeds that of P by unity. The maximum weight of a polynomial of degree d in $a_0, \ldots, a_p$ is pd, which is the weight of $a_p{}^d$. Since $O^{pd-w+1} S$ is of degree d and weight $pd + 1$, it is zero identically. Thus (11) for $r = pd - w + 1$ gives $O^{pd-w} S \equiv 0$. Then (11) for $r = pd - w$ gives $O^{pd-w-1} S \equiv 0$. Proceeding similarly, we get $OS \equiv 0$ and then $S \equiv 0$, contrary to hypothesis.

By Theorem 12 of §10, any covariant with the leader S is of order $n = pd - 2w$. Give it the notation

$$K = Sx^n + S_1 x^{n-1} y + S_2 x^{n-2} y^2 + \cdots + S_n y^n.$$

The result of applying the operator (7) to K is

$$(OS - S_1)x^n + (OS_1 - 2S_2)x^{n-1} y \\ + \cdots + (OS_{n-1} - nS_n)xy^{n-1} + OS_n y^n,$$

which is zero identically in x and y if and only if

(12) $\quad K = Sx^n + OSx^{n-1} y + \tfrac{1}{2} O^2 Sx^{n-2} y^2 + \cdots + \dfrac{1}{n!} O^n Sy^n,$

and $O^{n+1} S \equiv 0$. To show that the last condition is satisfied, note that (11) for $r = n + 1$ gives $\Omega O^{n+1} S \equiv 0$. Since $O^{n+1} S$ is of degree d and weight $W = w + n + 1 = pd - w + 1$ and is annihilated by Ω, it is a seminvariant by corollary 2 of §14. It is zero identically by the lemma since $pd - 2W = -(pd - 2w) - 2$ is negative. Hence (12) is annihilated by the operator (7).

In the result of applying operator (6) to (12), the coefficient of $x^{n-r} y^r$ is

$$\frac{1}{r!} \Omega O^r S - \frac{1}{(r-1)!}(n - r + 1) O^{r-1} S,$$

which is identically zero by (11). Hence (12) is a covariant by Theorem 1.

THEOREM 3. *There exists a unique covariant of a binary p-ic whose leader is any given seminvariant of the p-ic.*

17. Hilbert's theorem 4. *Any set S of forms in $x_1, \ldots, x_n$ contains a finite number of forms $F_1, \ldots, F_k$ such that every form of the set can be expressed as a linear combination of $F_1, \ldots, F_k$ with coefficients which are forms in $x_1, \ldots, x_n$, but are not necessarily in the set S.*

For $n = 1$, S is composed of certain forms $c_1 x^{e_1}, c_2 x^{e_2}, \ldots$. Let e_m be one of the minimum exponents and take $F_1 = c_m x^{e_m}$. Every form of the set S is the product of F_1 by a factor of type kx^e, where $e \geq 0$. Hence the theorem holds with $k = 1$ when $n = 1$.

To proceed by induction on n, let the theorem hold for every set of forms in $n - 1$ variables. To prove it for the set S, we may assume that S contains a form F_0 of (total) order r in which the coefficient of x_n^r is not zero. For, let $f(x_1, \ldots, x_n)$ be a form in S which actually involves x_n. Then there exist numbers $c_1, \ldots, c_n$ ($c_n \neq 0$) for which $c = f(c_1, \ldots c_n)$ is not zero. The transformation

$$x_i = y_i + c_i y_n \quad (i = 1, \ldots, n-1), \quad x_n = c_n y_n$$

§ 17] HILBERT'S THEOREM 31

is of determinant $c_n \neq 0$ and replaces f by a form $F_0(y_1, \ldots, y_n)$ in which the coefficient of $y_n{}^r$ is evidently c.

Let F be any form of the set S. Dividing F by F_0, we get $F \equiv F_0 P + R$, where R is a form whose order in x_n is $< r$. In R we segregate the terms whose order in x_n is exactly $r - 1$ and have

$$F \equiv F_0 P + M x_n{}^{r-1} + N,$$

where M is a form in $x_1, \ldots, x_{n-1}$, while N is a form in $x_1, \ldots, x_n$ whose order in x_n is $\leq r - 2$. Each F uniquely determines an M.

Since the theorem is true by hypothesis for the set of forms M, there exist a finite number $M_1, \ldots, M_l$ of them (corresponding to certain forms $F_1, \ldots, F_l$ of S), such that every M can be expressed as

$$M \equiv f_1 M_1 + \cdots + f_l M_l,$$

where the f_i are forms in $x_1, \ldots, x_{n-1}$. Then

$$F \equiv F_0 P + N + x_n{}^{r-1} \sum_{i=1}^{l} f_i M_i, \quad F_i \equiv F_0 P_i + M_i x_n{}^{r-1} + N_i,$$

whence

$$F \equiv F_0 P' + \sum f_i F_i + R', \quad P' = P - \sum f_i P_i,$$
$$R' = N - \sum f_i N_i.$$

Each exponent of x_n in R' is $\leq r - 2$; we segregate its terms for which that exponent is exactly $r - 2$ and have

$$F = F_0 P' + \sum_{i=1}^{l} f_i F_i + M' x_n{}^{r-2} + N',$$

where M' is a form in $x_1, \ldots, x_{n-1}$, and N' is a form in $x_1, \ldots, x_n$ whose order in x_n is $\leq r - 3$.

Since the theorem holds for the set of forms M', each is a linear combination of $M'_1, \ldots, M'_m$, corresponding to certain forms $F_{l+1}, \ldots, F_{l+m}$ of S. As before, F differs from a linear combination of $F_0, \ldots, F_{l+m}$ by $M'' x_n{}^{r-3} + N''$, where M'' is a form in $x_1, \ldots, x_{n-1}$, and N'' is a form whose order in x_n is $\leq r - 4$. Proceeding similarly, we see that F differs from a linear combination of certain forms $F_0, \ldots, F_t$ of S by a form lacking x_n. One more step leads to the theorem.

18. Finiteness of a fundamental system of covariants. For simplicity, we first consider only invariants of a single binary form f with the coefficients $a_0, \ldots, a_p$. By Theorem 8 of §8, every invariant of f is homogeneous in the a's. By the preceding theorem, there is a finite number of these invariants $I_1, \ldots, I_m$ in terms of which every invariant I of F is expressible linearly:

(13) $$I \equiv E_1 I_1 + \cdots + E_m I_m.$$

Let I and I_j be of degrees d and d_j, respectively. By the corollary at the end of §10, their weights are $\frac{1}{2}pd$ and $\frac{1}{2}pd_j$. Hence E_j is of degree $D_j = d - d_j$ and weight $\frac{1}{2}pD_j$.

We shall next obtain a differential operator D such that DE_j is an invariant. Apply the operator O^{r-2} to (10); we get

$$O^{r-1}\, \Omega^r\, S - O^{r-2}\, \Omega^r\, OS \equiv r(-n-r+1) O^{r-2}\, \Omega^{r-1}\, S.$$

In the second term replace ΩOS by its value $O\Omega S + nS$ from (8). Thus

$$O^{r-1}\, \Omega^r\, S \equiv O^{r-2}\, \Omega^{r-1}[O\Omega - (r-1)(n+r)]S.$$

Since $O\Omega S$ has the same degree and weight as S, we may employ the formula derived from this by replacing r by $r - 1$ to eliminate $O^{r-2}\, \Omega^{r-1}$, and by repetitions of this process evidently obtain

$$O^{r-1}\, \Omega^r\, S \equiv \Omega[O\Omega - (n+2)][O\Omega - 2(n+3)] \cdots [O\Omega - (r-1)(n+r)]S,$$

which we may also establish by induction. Let S be of degree ∂ and weight $w = \frac{1}{2}p\partial$, so that its n in (8) is zero. Take $r = w + 1$, apply $\Omega^{w+1} S \equiv 0$, and divide by $(-2)(-2 \cdot 3) \ldots$; we get $0 \equiv \Omega DS$, where

$$(14) \quad D = \left[1 - \frac{O\Omega}{1 \cdot 2}\right]\left[1 - \frac{O\Omega}{2 \cdot 3}\right] \cdots \left[1 - \frac{O\Omega}{w(w+1)}\right].$$

Since DS is annihilated by Ω and is of constant degree ∂ and constant weight $w = \frac{1}{2}p\partial$, it is a seminvariant by Corollary 2 of §14. Being the leader of a covariant of order $n = 0$, it is an invariant. Since the invariants I and I_j are annihilated by both Ω and O, they are unaltered by the operator D. These results hold also if we annex factors $g = 1 - cO\Omega$ at the right of (14), since gS has the same degree and weight as S. Hence if we take w to be the maximum of the weights of the E_j, and operate on (13) by D, we get

$$I \equiv J_1 I_1 + \cdots + J_m I_m, \quad J_j = DE_j = \text{invariant}.$$

Since each J_j is of the form (13),

$$J_j \equiv \sum_{k=1}^{m} e_{jk} I_k, \quad I \equiv \sum_{j,\,k=1}^{m} e_{jk} I_j I_k.$$

Applying to the last an operator D with w sufficiently large, we get

$$I \equiv \sum L_{jk} I_j I_k,$$

where the L_{jk} are invariants of f. Since there is a reduction of degree at each step of the process, we ultimately obtain an expression for I as a polynomial in $I_1, \ldots, I_m$ with numerical coefficients. Hence $I_1, \ldots, I_m$ form a fundamental system of invariants of f.

Exercise

Extend the finiteness proof to invariants of a system of binary forms $a_0 x^{p_1} + \ldots, \ b_0 x^{p_2} + \ldots, \ c_0 x^{p_3} + \ldots, \ \ldots$ First prove as in §14 that an invariant is annihilated by

$$\Sigma\Omega = a_0 \frac{\partial}{\partial a_1} + \cdots + b_0 \frac{\partial}{\partial b_1} + \cdots + p_2 b_{p_2-1} \frac{\partial}{\partial b_{p_2}} + c_0 \frac{\partial}{\partial c_1} + \cdots$$

and by $\Sigma\, O$. Next, if S is a polynomial in the a_i, b_i, ... which is of constant degree d_1 in the a's, of constant degree d_2 in the b's, ..., and of total weight w in the a's, b's, ... taken collectively, show that (8) and (10) hold if we replace O by $\Sigma\, O$, Ω^r by $(\Sigma\, \Omega)^r$, and n by $\Sigma\, p_i\, d_i - 2w$. Make the like replacements in (14).

The finiteness proof is extended to covariants by means of the

LEMMA. *The set of all homogeneous covariants of the binary forms $f_1, \ldots, f_k$ is identical with the set of forms derived from the invariants I (homogeneous in X, Y) of $f_1, \ldots, f_k$ and $L \equiv Yx - Xy$ by replacing X by x and Y by y in each I.*

To simplify the notations, let $a, b, \ldots$ denote the coefficients of $f_1, \ldots, f_k$ arranged in any chosen order. Let $A, B, \ldots$ denote the corresponding coefficients of the forms obtained by the transformation

$$x = pu + qv, \quad y = ru + sv, \quad D = ps - qr \neq 0.$$

This replaces L by $Vu - Uv$, in which

$$U = sX - qY, \quad V = -rX + pY.$$

Solving these two equations, we get

$$DX = pU + qV, \quad DY = rU + sV.$$

Let $I = I(a, b, \ldots; X, Y)$ be of index l. Then

$$I(A, B, \ldots; U, V) = D^l\, I = D^{l-n}\, I(a, b, \ldots; DX, DY),$$

since I is of constant order n in X, Y. Hence

$$I(A, B, \ldots; U, V)$$
$$\equiv D^{l-n}\, I(a, b, \ldots; pU + qV, rU + sV),$$

identically in U, V. We may replace the latter by u, v and get

$$I(A, B, \ldots; u, v) = D^{l-n}\, I(a, b, \ldots; x, y).$$

This proves that $I(a, b, \ldots; x, y)$ is a covariant of $f_1, \ldots, f_k$ of order n and index $l - n$. The argument may be reversed.

Since every covariant is a sum of covariants each homogeneous in the variables (§8), we have completed the proof of

THEOREM 5. *There exists a fundamental system of covariants of every set of binary forms.*

The first proof of this classic theorem was made by Gordan by use of the symbolic notation. Cayley had earlier come to the contrary conclusion that the fundamental system for a binary quintic is infinite, due to an error relating to the independence of the syzygies between the covariants. For a binary form of order <5, a finite fundamental system of covariants was constructed in §12.

19. Finiteness of syzygies. Let $I_1, \ldots, I_m$ be a fundamental system of invariants of the binary forms $f_1, \ldots, f_r$. Let $S(z_1, \ldots, z_m)$ be a polynomial with numerical coefficients such that $S(I_1, \ldots, I_m)$ is zero identically in the coefficients of the f's. Then $S(I_1, \ldots, I_m) = 0$ is a syzygy between the invariants (see the examples in §12).

By means of a new variable z_{m+1}, we may convert $S(z_1, \ldots, z_m)$ into a homogeneous form H by multiplying a suitable power of z_{m+1} into all terms not of the maximum degree. By Hilbert's theorem, the forms H are all expressible linearly in terms of a finite number $H_1, \ldots, H_k$ of them. Take $z_{m+1} = 1$. Thus

(15) $$S \equiv P_1 S_1 + \cdots + P_k S_k,$$

where $P_1, \ldots, P_k$ are polynomials in $z_1, \ldots, z_m$. In this identity in $z_1, \ldots, z_m$, we may take $z_1 = I_1, \ldots, z_m = I_m$.

THEOREM 6. *There is a finite number of syzygies $S_1 = 0, \ldots, S_k = 0$, such that every syzygy $S = 0$ implies a relation* (15) *in which $P_1, \ldots, P_k$ are now invariants.*

In particular, every syzygy is a consequence of $S_1 = 0, \ldots, S_k = 0$. By the Lemma in §18, the proof holds also for syzygies between covariants.

20. Canonical form of a binary form of odd order. In §7 we saw that any binary cubic whose discriminant is not zero is the sum of the cubes of two linear forms which are the factors of the Hessian. We shall prove an analogous theorem for a form f of any odd order $2n - 1$. Let f have the notation (1) with $p = 2n - 1$. We seek constants p_i, r_i such that

(16) $\qquad f \equiv p_1(x + r_1 y)^{2n-1} + \cdots + p_n(x + r_n y)^{2n-1},$

identically in x, y. The conditions are evidently

(17) $\qquad a_i = p_1 r_1^i + \cdots + p_n r_n^i \qquad (i = 0, 1, \ldots, 2n - 1).$

We can determine solutions $q_n, \ldots, q_1$ of the n linear equations

(18) $\qquad a_s q_n + a_{s+1} q_{n-1} + \cdots + a_{s+n-1} q_1 + a_{s+n} = 0$
$\qquad\qquad\qquad\qquad\qquad (s = 0, 1, \ldots, n - 1)$

if the determinant of their coefficients is not zero. Let $r_1, \ldots, r_n$ be the roots of

(19) $\qquad z^n + q_1 z^{n-1} + \cdots + q_{n-1} z + q_n = 0.$

If $r_1, \ldots, r_n$ are distinct, conditions (17) with $i = 0, 1, \ldots, n - 1$ uniquely determine $p_1, \ldots, p_n$. We shall prove that the remaining n conditions are then satisfied. To proceed by induction on s, we assume that the conditions (17) hold when $i < s + n$ and prove that the condition with $i = s + n$ is then satisfied. Multiply the condition with $i = s$ by q_n, that with $i = s + 1$ by $q_{n-1}, \ldots,$ that with $i = s + n - 1$ by q_1, that with $i = s + n$ by unity, and add. The coefficient of p_k is

$$r_k{}^s q_n + r_k{}^{s+1} q_{n-1} + \cdots + r_k{}^{s+n-1} q_1 + r_k{}^{s+n},$$

which is the product of $r_k{}^s$ by the zero value of the polynomial (19) for $z = r_k$. The new left member is evidently (18).

The equation having the roots $r_1, \ldots, r_n$ is evidently obtained by eliminating $q_1, \ldots, q_n$ from (18) and (19). We prefer the homogeneous equation in x and y whose linear factors are $x + r_i y$ $(i = 1, \ldots, n)$. It is found by taking the determinant of the coefficients of $q_n, \ldots, q_1, 1$ in

(19') $\quad y^n q_n - xy^{n-1} q_{n-1} + x^2 y^{n-2} q_{n-2}$
$\qquad\qquad - \cdots + (-x)^{n-1} y\, q_1 + (-x)^n = 0$

and (18), and hence is

$$\begin{vmatrix} y^n & -xy^{n-1} & x^2 y^{n-2} & \cdots & (-x)^{n-1} y & (-x)^n \\ a_0 & a_1 & a_2 & \cdots & a_{n-1} & a_n \\ a_1 & a_2 & a_3 & \cdots & a_n & a_{n+1} \\ \cdots & & & & & \\ a_{n-1} & a_n & a_{n+1} & \cdots & a_{2n-2} & a_{2n-1} \end{vmatrix} = 0.$$

This determinant is called the *canonizant* C of f.

THEOREM 7. *Every sufficiently general binary form of odd order $2n - 1$ is a sum of $(2n - 1)$th powers of n linear forms.*

EXERCISES

1. For $n = 3$, the coefficient of x^3 in the canonizant C of the quintic f is the invariant J of the quartic $q = a_0 x^4 + 4a_1 x^3 y + \ldots$. In particular, J is a seminvariant of q and hence is annihilated by the operator Ω in (6) with $p = 4$. Since J lacks a_5 it is therefore annihilated by Ω with $p = 5$ and hence is a seminvariant of f.

2. Show that J is a seminvariant of every f with $p \geq 4$ since it is a rational function of the seminvariants a_0, S_2, S_3, S_4. See (21) of §12, and §10.

3. There exists a covariant of order 3 of the quintic f with the leader J. It coincides with C since the canonizant of the form derived from f by any linear transformation T has as linear factors the functions derived from the linear factors of C by T.

4. For $n = 3$, the canonizant C of f is the minor of y^3 in Δ:

$$\Delta = \begin{vmatrix} y^3 & 0 & 0 & 0 \\ a_0 & a_0 x + a_1 y & a_1 x + a_2 y & a_2 x + a_3 y \\ a_1 & a_1 x + a_2 y & a_2 x + a_3 y & a_3 x + a_4 y \\ a_2 & a_2 x + a_3 y & a_3 x + a_4 y & a_4 x + a_5 y \end{vmatrix}, \quad D = \begin{vmatrix} 1 & x & 0 & 0 \\ 0 & y & x & 0 \\ 0 & 0 & y & x \\ 0 & 0 & 0 & y \end{vmatrix}$$

For, the product of C by $D = y^3$, rows by columns, is Δ.

5. If a seminvariant S of the quintic f lacks a_5, it is a seminvariant of the quartic q. If the weight of S is double its degree, S is an invariant of q.

6. Find the determinant whose vanishing is a necessary and sufficient condition that a binary form f of order $2n$ be a sum of the $2n$th powers of n linear forms.

Hint: Employ (18) also for $s = n$.

21. Further topics on covariants. Reference will be made to elementary introductions in English to the following subjects:

Seminvariants and covariants as functions of the roots.[1]

Symbolic notation; geometrical applications.[2]

Canonical form of the binary sextic and octavic.[3]

Functionally complete system of covariants of a binary form, found from a complete system of three linear partial differential equations.[4]

Theory of covariants under linear transformations with integral coefficients taken modulo p, a prime.[5]

The coefficients of a unique canonical form are invariants.[6]

Interesting applications of invariants of a quintic form are given in the elementary papers cited in §112.

[1] Dickson, Algebraic Invariants, Wiley and Sons, 1914, pp. 53–58.

[2] *Ibid.*, pp. 63–97. The Algebra of Invariants by Grace and Young is very difficult for students, except possibly the chapters on geometry. To readers familiar with German are recommended Gordan, Vorlesungen über Invariantentheorie, 1887; Clebsch-Lindemann, Vorlesungen über Geometrie, I, 1876; II, 1891; and the up-to-date book on Invariantentheorie by Weitzenböck, Groningen, 1923. The specialist will need Meyer's report on invariants in Jahresbericht der Deutschen Math.-Vereinigung, 1, 1890–2, 79–292.

[3] Elliott, Algebra of Quantics, 1895, 294–9; ed. 2, 1913, 285–90.

[4] Dickson, Annals of Math., 25, 1924, 369–76. We may start with (18), p. 373, noting that uf, vf, and wf are Of, Ωf, and $(\Omega O - O\Omega)f$ of our text.

[5] Dickson, On Invariants and the Theory of Numbers, the Madison Colloquium of the Amer. Math. Soc., 1914. Cf. Glenn, Treatise on the Theory of Invariants, 1915, 175–208. Dickson, History of the Theory of Numbers, III, 1923, 293–301.

[6] Wilczynski, Proc. National Acad. Sciences, 4, 1918, 300–5.

Chapter III

MATRICES, BILINEAR FORMS, LINEAR EQUATIONS

Chapters III–VI, which are independent of I–II, give a new exposition of the subject usually called higher algebra. We first develop Cayley's calculus of matrices, and the essentially equivalent subject of bilinear forms. The main theorems on the solution of systems of linear equations are not presupposed, but are deduced as corollaries.

22. Linear forms, matrices, linear transformations. Any homogeneous linear function, such as $7x - 3y$, is called a *linear form*. Consider m linear forms

$$(1) \quad \begin{cases} l_1 = a_{11} y_1 + a_{12} y_2 + \cdots + a_{1n} y_n, \\ \cdots\cdots\cdots\cdots\cdots\cdots\cdots\cdots\cdots\cdots\cdots \\ l_m = a_{m1} y_1 + a_{m2} y_2 + \cdots + a_{mn} y_n \end{cases}$$

in the n variables $y_1, \ldots, y_n$. Write the coefficients of these successive linear forms in the successive rows of a table

$$(2) \quad A = \begin{bmatrix} a_{11} & a_{12} & \cdots & a_{1n} \\ \cdots & \cdots & \cdots & \cdots \\ a_{m1} & a_{m2} & \cdots & a_{mn} \end{bmatrix}.$$

Such a rectangular table composed of mn numbers a_{ij} arranged in m rows with n numbers in each row is called a *matrix* with m rows and n columns, or briefly an $m \times n$ matrix. The mn numbers a_{ij} are called the *elements* of the matrix. We shall speak of A as the matrix of (the coefficients of) the linear forms (1).

Examples of one-rowed matrices are the notations (x, y) and (x, y, z) for points in a plane and in space, the elements of these matrices being the coordinates of the points.

Matrix (2) is said to be *equal* to the $m \times n$ matrix

$$\begin{bmatrix} d_{11} & d_{12} & \cdots & d_{1n} \\ \cdots\cdots\cdots\cdots\cdots \\ d_{m1} & d_{m2} & \cdots & d_{mn} \end{bmatrix}$$

if and only if corresponding elements are equal:

$$a_{11} = d_{11}, \quad a_{12} = d_{12}, \ldots, \quad a_{mn} = d_{mn}.$$

We shall also write (1) and (2) in the compact notations

(1') $\quad l_i = \sum_{j=1}^{n} a_{ij} y_j \quad (i = 1, \ldots, m),$

(2') $\quad A = (a_{ij}) \quad (i = 1, \ldots, m; j = 1, \ldots, n),$

in which we exhibit the element a_{ij} which lies in the ith row and the jth column of the matrix (2).

Suppose that the n variables y_j are expressible linearly in terms of s new variables $z_1, \ldots, z_s$:

(3) $\quad y_j = \sum_{k=1}^{s} b_{jk} z_k \quad (j = 1, \ldots, n).$

Inserting these values into (1'), we get

(4) $\quad l_i = \sum_{j=1}^{n} \sum_{k=1}^{s} a_{ij} b_{jk} z_k = \sum_{k=1}^{s} p_{ik} z_k \quad (i = 1, \ldots, m),$

where

(5) $\quad p_{ik} = \sum_{j=1}^{n} a_{ij} b_{jk} \quad (i = 1, \ldots, m; k = 1, \ldots, s).$

Let $B = (b_{jk})$ and $P = (p_{ik})$ denote the matrices of the linear forms (3) and (4) respectively:

§ 22]   LINEAR FORMS, MATRICES   41

(6) $$B = \begin{pmatrix} b_{11} & \cdots & b_{1s} \\ \cdots & \cdots & \cdots \\ b_{n1} & \cdots & b_{ns} \end{pmatrix}, \qquad P = \begin{pmatrix} p_{11} & \cdots & p_{1s} \\ \cdots & \cdots & \cdots \\ p_{m1} & \cdots & p_{ms} \end{pmatrix}.$$

We shall call P the *product* of A and B, taken in that order, and shall write $P = AB$. In view of (5), we therefore have the following

Rule of multiplication of matrices: The element in the ith row and kth column of the product of a matrix A with n columns by a matrix B with n rows is the sum of the products of the successive elements of the ith row of A by the corresponding elements of the kth column of B.

For example,

$$\begin{pmatrix} a & b \\ c & d \end{pmatrix} \begin{pmatrix} \alpha & \beta \\ \gamma & \partial \end{pmatrix} = \begin{pmatrix} a\alpha + b\gamma & a\beta + b\partial \\ c\alpha + d\gamma & c\beta + d\partial \end{pmatrix}.$$

By a *linear transformation* we shall mean a system of n linear homogeneous equations

(7) $$y_j = \sum_{k=1}^{n} b_{jk} z_k \qquad (j = 1, \ldots, n),$$

which express the variables $y_1, \ldots, y_n$ linearly in terms of the same number of variables $z_1, \ldots, z_n$. This transformation is said to have the matrix (b_{jk}). Taking $s = n$ in (3)–(6), we shall say that the transformation (7) replaces the system of linear forms (1) by the system of linear forms (4). This proves

THEOREM 1. *A linear transformation with the matrix B replaces a system of linear forms with the matrix A by a system of linear forms with the matrix AB.*

When $m = n$, equations (1′) define a linear transformation

(8) $$l_i = \sum_{j=1}^{n} a_{ij} y_j \qquad (i = 1, \ldots, n),$$

which expresses the variables $l_1, \ldots, l_n$ linearly in terms of the variables $y_1, \ldots, y_n$. We saw that the elimination of $y_1, \ldots, y_n$ between equations (1′) and (3) gave (4) and (5), which, in our special case $m = n = s$, become

$$l_i = \sum_{k=1}^{n} p_{ik} z_k, \qquad p_{ik} = \sum_{j=1}^{n} a_{ij} b_{jk} \qquad (i, k = 1, \ldots, n).$$

The resulting linear transformation, which expresses $l_1, \ldots, l_n$ linearly in terms of $z_1, \ldots, z_n$, and has the matrix AB, is called the *product* of the linear transformation (8) with the matrix A by the linear transformation (7) with the matrix B.

The one-to-one correspondence between linear transformations and square matrices is therefore preserved under multiplication. For this reason a linear transformation is often denoted by its matrix when the particular choice of the letters used to denote the variables is immaterial.

The determinant $|a_{ij}|$ having the same n^2 elements as the matrix of a linear transformation (8) is called the *determinant of the transformation*. Since the above rule of multiplication of square matrices is one of the rules of multiplication of determinants, we conclude that *the determinant of the product of two linear transformations is equal to the product of their determinants.*

A linear transformation is called *singular* if its determinant is zero, otherwise *non-singular*.

23. Inverse, identity, scalar, adjoint. Let a denote the determinant of transformation (8). If $a \neq 0$, equations (8) have the unique solution

$$(9) \qquad y_j = \sum_{k=1}^{n} \frac{A_{kj}}{a} l_k \qquad (j = 1, \ldots, n),$$

where A_{kj} denotes the cofactor of a_{kj} in a. Equations (9) define a linear transformation which expresses the y's linearly in terms of

§ 23]   INVERSE, IDENTITY, ADJOINT   43

the l's; it is called the *inverse* of transformation (8). Since it is immaterial what letters are used to denote variables, let us replace each l_k by z_k. Then the product of (8) by its inverse

(9') $$y_j = \sum_{k=1}^{n} \frac{A_{kj}}{a} z_k \qquad (j = 1, \ldots, n)$$

is evidently the *identity transformation* I defined by

(10) $$l_1 = z_1, \ldots, l_n = z_n.$$

The inverse of (9) is evidently the initial transformation (8). Note that $I = S_1$ if S_t denotes

(11) $$u_1 = tv_1, \ldots, u_n = tv_n.$$

The product of (8) by S_t in either order is

$$l_i = \sum_{j=1}^{n} t a_{ij} y_j \qquad (i = 1, \ldots, n),$$

whose matrix will be denoted by either tA or At, being obtained by multiplying every element of A by t. The matrix of (11) will be denoted by S_t and called a *scalar* matrix; its diagonal elements are t and all remaining elements are zero. In particular, the matrix of (10) is denoted by I and is called the *identity* matrix. Thus $S_1 = I$. Our results give

(12) $$S_t A = A S_t = tA, \qquad IA = AI = A.$$

Hence in products we may replace a scalar matrix factor S_t by the number t. In particular, we may suppress a factor I.

A matrix all of whose elements are zero is called a *zero* matrix and designated by 0. By (12) with $t = 0$, we have $0A = A0 = 0$ for every matrix A.

The matrix of (9′) is designated by

(13) $$A^{-1} = \begin{bmatrix} A_{11}/a & \cdots & A_{n1}/a \\ \cdots\cdots\cdots\cdots\cdots \\ A_{1n}/a & \cdots & A_{nn}/a \end{bmatrix},$$

and called the *inverse of A*. By the above results,

(14) $$AA^{-1} = I, \qquad A^{-1}A = I.$$

If we suppress the denominators a in (13), we get

(15) $$\text{Adj. } A = \begin{bmatrix} A_{11} & \cdots & A_{n1} \\ \cdots\cdots\cdots\cdots \\ A_{1n} & \cdots & A_{nn} \end{bmatrix},$$

which is called the *adjoint* of matrix A. Hence the element A_{ji} in the ith row and jth column of the adjoint of A is the cofactor of the element a_{ji} in the jth row and ith column of A.

In case the determinant a of A is not zero, we have Adj. $A = a A^{-1}$ by definition; then (14) implies

(16) $$A(\text{Adj. } A) = aI, \qquad (\text{Adj. } A)A = aI.$$

This result (16) holds true also when $a = 0$, as shown by consideration of continuity or by direct multiplication.

24. Associative and distributive laws for matrices. Let $A = (a_{ij})$, $B = (b_{ij})$, $C = (c_{ij})$ be any $m \times n$, $n \times s$, $s \times t$ matrices, respectively. By (5), the element $[ik]$ in the ith row and kth column of the product AB is $\sum_j a_{ij} b_{jk}$. By the same rule, the element $[il]$ of $(AB)C$ is

(17) $$\sum_{k=1}^{s} \Big(\sum_{j=1}^{n} a_{ij} b_{jk} \Big) c_{kl}.$$

The element $[jl]$ of BC is $\sum_k b_{jk} c_{kl}$. Hence the element $[il]$ of $A(BC)$ is

ASSOCIATIVE, DISTRIBUTIVE LAWS

$$\sum_{j=1}^{n} a_{ij} \left(\sum_{k=1}^{s} b_{jk} c_{kl} \right).$$

Since this is equal to the sum (17), we have

THEOREM 2. *Multiplication of matrices is associative,*[1] *i.e.,*

(18) $\qquad (AB)C = A(BC).$

Either of these products may therefore be denoted by ABC without ambiguity. It follows that $ABC \cdot D$, $AB \cdot CD$ and $A \cdot BCD$ are equal and may be denoted by $ABCD$. Similarly for the product of any number of matrices taken in a prescribed order. The case of equal factors shows that A^e is uniquely defined when the exponent e is any positive integer.

The sum (or difference) of two $m \times n$ matrices is defined to be the $m \times n$ matrix each of whose elements is the sum (or difference) of the corresponding elements of the given matrices. For example,

$$\begin{pmatrix} a & b \\ c & d \end{pmatrix} \pm \begin{pmatrix} \alpha & \beta \\ \gamma & \partial \end{pmatrix} = \begin{pmatrix} a \pm \alpha & b \pm \beta \\ c \pm \gamma & d \pm \partial \end{pmatrix}.$$

It follows at once that

(19) $\quad \begin{cases} A + B = B + A, \ (A + B) + C = A + (B + C), \\ tA + tB = t(A + B), \ tA + uA = (t + u)A. \end{cases}$

The element $[ik]$ of $A(B + C)$ is $\sum_j a_{ij}(b_{jk} + c_{jk})$, which is the sum of the elements $[ik]$ of AB and AC. In this manner, we obtain

THEOREM 3. *Multiplication of matrices is distributive, i.e.,*

(20) $A(B \pm C) = AB \pm AC, \qquad (B \pm C)A = BA \pm CA.$

---

[1] Another proof follows from §1.

In view of the associative law, *the inverse of any product of matrices is the product of their inverses taken in reverse order*. For example,

$$ABC \cdot C^{-1} B^{-1} A^{-1} = A \cdot BB^{-1} \cdot A^{-1} = AA^{-1} = I.$$

If A is non-singular, $AX = B$ has the unique solution $X = A^{-1} B$, while $YA = B$ has the solution $Y = BA^{-1}$. Since matrix multiplication is not commutative in general, X and Y are usually distinct, so that there are two kinds of division of B by A.

Exercises

1. Scalar matrices are the only ones commutative with every $n \times n$ matrix.

2. Show that A^{-1} is the only matrix whose products with A in both orders gives the identity matrix.

3. By multiplication of the matrices

$$I = \begin{pmatrix} 1 & 0 \\ 0 & 1 \end{pmatrix}, \quad i = \begin{pmatrix} \sqrt{-1} & 0 \\ 0 & -\sqrt{-1} \end{pmatrix}, \quad j = \begin{pmatrix} 0 & 1 \\ -1 & 0 \end{pmatrix}, \quad k = \begin{pmatrix} 0 & \sqrt{-1} \\ \sqrt{-1} & 0 \end{pmatrix},$$

verify that $i^2 = j^2 = k^2 = -I$, $ij = k$. By the associative law,

$$ik = iij = -j, \quad kj = ijj = -i, \quad ki = k(-kj) = j,$$
$$ji = kii = -k, \quad jk = j(-ji) = i.$$

Any linear combination of I, i, j, k with numerical coefficients is called a *quaternion*. The two quaternions

$$q = aI + bi + cj + dk, \qquad q' = aI - bi - cj - dk$$

are called *conjugates*. Verify that

$$q = \begin{pmatrix} x & y \\ -\bar{y} & \bar{x} \end{pmatrix}, \qquad q' = \begin{pmatrix} \bar{x} & -y \\ \bar{y} & x \end{pmatrix}, \qquad \begin{aligned} x &= a + b\sqrt{-1}, \\ y &= c + d\sqrt{-1}, \end{aligned}$$

and that $qq' = q'q = nI$, where $n = x\bar{x} + y\bar{y}$ is called the *norm* of q and written $n(q)$.

4. The conjugate of the product of two quaternions is equal to the product of their conjugates taken in reverse order. For,

$$p = \begin{pmatrix} u & v \\ -\bar{v} & \bar{u} \end{pmatrix}, \quad qp = \begin{pmatrix} r & s \\ -\bar{s} & \bar{r} \end{pmatrix}, \quad p'q' = \begin{pmatrix} \bar{r} & -s \\ \bar{s} & r \end{pmatrix} = (qp)',$$

where $r = xu - y\bar{v}$, $s = xv + y\bar{u}$. To find the norm of qp, note that

$$qp \cdot p'q' = q \cdot n(p)I \cdot q' = qq' \cdot n(p)I = n(q)n(p)I.$$

Hence the norm of a product is the product of the norms of the factors.

5. In matrices (2) and (6), let the a_{ij}, b_{ij}, p_{ij} be themselves matrices, such that those appearing in any row have the same number of rows, while those appearing in any column have the same number of columns. Prove that (5) holds also here. If each $b_{ij} = 0$ when $i \neq j$, find the two products of (a_{ij}) and (b_{ij}).

25. Characteristic equation. Let A be an n-rowed square matrix whose elements are independent of the variable λ. Let I be the n-rowed identity matrix. The matrix $A - \lambda I$ is called the *characteristic matrix* of A; it may be obtained by subtracting λ from each diagonal element of A. The *characteristic determinant* of A is the determinant of $A - \lambda I$ and is a polynomial in λ of degree n:

(21) $\quad \phi(\lambda) \equiv a_n \lambda^n + a_{n-1} \lambda^{n-1} + \cdots + a_1 \lambda + a_0, \quad a_n = (-1)^n.$

The equation $\phi(\lambda) = 0$ is called the *characteristic equation* of A.

For example,

$$A = \begin{pmatrix} a & b \\ c & d \end{pmatrix},$$

(22)
$$\phi(\lambda) = \begin{vmatrix} a - \lambda & b \\ c & d - \lambda \end{vmatrix} = \lambda^2 - (a+d)\lambda + ad - bc.$$

If in any polynomial $f(\lambda) = \sum c_i \lambda^i$ we substitute the matrix A for λ and multiply the constant term c_0 by I, we obtain a matrix denoted by $f(A)$. In the special case (22), we get

$$\phi(A) = A^2 - (a+d)A + (ad-bc)I,$$

which is readily verified to be equal to matrix 0. This result is the case $n = 2$ of the important

THEOREM 4. *Any square matrix satisfies its characteristic equation.*

Let (21) be the characteristic determinant of A. We are to prove that $\phi(A) = 0$. Since the elements of $A - \lambda I$ are linear functions of λ, and the elements of its adjoint C are $(n-1)$-rowed determinants, they are polynomials in λ of degree $\leq n-1$. If the element in the ith row and jth column of C is $\sum_k c_{ijk} \lambda^k$, then

$$C = \sum_{k=0}^{n-1} C_k \lambda^k, \qquad C_k = (c_{ijk}) \qquad (i,j = 1, \ldots, n).$$

Applying (16) with A replaced by $A - \lambda I$, we have

$$(A - \lambda I)C = \phi(\lambda)I.$$

Hence

$$A \sum_{k=0}^{n-1} C_k \lambda^k - \lambda \sum_{k=0}^{n-1} C_k \lambda^k = \sum_{k=0}^{n} a_k \lambda^k I.$$

Equating the terms free of λ and the coefficients of $\lambda, \lambda^2, \ldots, \lambda^{n-1}, \lambda^n$, we get

$$\begin{aligned} AC_0 &= a_0 I, \\ AC_1 - C_0 &= a_1 I, \\ AC_2 - C_1 &= a_2 I, \\ &\cdots\cdots\cdots \\ AC_{n-1} - C_{n-2} &= a_{n-1} I, \\ -C_{n-1} &= a_n I. \end{aligned}$$

Multiply these equations on the left by $I, A, A^2, \ldots, A^{n-1}, A^n$ respectively and add; we get

$$0 = a_0 I + a_1 A + a_2 A^2 + \cdots + a_{n-1} A^{n-1} + a_n A^n \equiv \phi(A).$$

26. Rank of a matrix.

Every matrix M having more than one element contains other matrices obtained from M by deleting certain rows or columns or both. In particular, it contains certain square matrices. The determinants of these square matrices are called the *determinants of* M.

A matrix is said to be of *rank* r if it contains at least one r-rowed determinant which is not zero, while all its determinants of order higher than r are zero. The zero matrix all of whose elements are zero is said to be of rank 0.

For example, the rank of the matrix

$$\begin{bmatrix} a & b & c & d & e \\ p & q & r & s & t \\ a & b & c & d & e \\ p & q & r & s & t \end{bmatrix}$$

is 2 or less since every three-rowed determinant has two rows alike and is zero. The rank is 2 if a, b, c, d, e are not proportional to p, q, r, s, t, so that a two-rowed determinant is not zero. The rank is 1 if they are proportional and not all zero.

By the rank of a determinant is meant the rank of its matrix.

THEOREM 5. *If A is any matrix with n columns and B is any matrix with n rows, any t-rowed determinant D of matrix AB is equal to a sum of terms each a product of a t-rowed determinant of A by a t-rowed determinant of B.*

Let $A = (a_{ij})$, $B = (b_{ij})$, $P = AB = (p_{ik})$ be given by (2), (5), (6). For the case $n = 3$, $t = 2$, the theorem is illustrated by the formula

$$\begin{vmatrix} p_{12} & p_{13} \\ p_{22} & p_{23} \end{vmatrix} = \begin{vmatrix} a_{11}b_{12}+a_{12}b_{22}+a_{13}b_{32} & a_{11}b_{13}+a_{12}b_{23}+a_{13}b_{33} \\ a_{21}b_{12}+a_{22}b_{22}+a_{23}b_{32} & a_{21}b_{13}+a_{22}b_{23}+a_{23}b_{33} \end{vmatrix}$$

$$= \begin{vmatrix} a_{11} & a_{12} \\ a_{21} & a_{22} \end{vmatrix} \cdot \begin{vmatrix} b_{12} & b_{13} \\ b_{22} & b_{23} \end{vmatrix} + \begin{vmatrix} a_{11} & a_{13} \\ a_{21} & a_{23} \end{vmatrix} \cdot \begin{vmatrix} b_{12} & b_{13} \\ b_{32} & b_{33} \end{vmatrix} + \begin{vmatrix} a_{12} & a_{13} \\ a_{22} & a_{23} \end{vmatrix} \cdot \begin{vmatrix} b_{22} & b_{23} \\ b_{32} & b_{33} \end{vmatrix}.$$

Any t-rowed determinant D of P is of the form

$$D = \begin{vmatrix} p_{i_1 k_1} & \cdots & p_{i_1 k_t} \\ \cdots & \cdots & \cdots \\ p_{i_t k_1} & \cdots & p_{i_t k_t} \end{vmatrix} = \begin{vmatrix} \sum_{j_1=1}^{n} a_{i_1 j_1} b_{j_1 k_1} & \cdots & \sum_{j_t=1}^{n} a_{i_1 j_t} b_{j_t k_t} \\ \cdots & \cdots & \cdots \\ \sum_{j_1=1}^{n} a_{i_t j_1} b_{j_1 k_1} & \cdots & \sum_{j_t=1}^{n} a_{i_t j_t} b_{j_t k_t} \end{vmatrix},$$

where $i_1 < i_2 < \cdots < i_t$, $k_1 < k_2 < \cdots < k_t$. Since each element of the first column is a sum of n terms, D is equal to the sum of n determinants of which the j_1th has as elements of the first column the j_1th terms of those sums and has as elements of the remaining columns the elements of D. In other words, we may remove the summation sign from the first column and place it in front of the symbol of the resulting determinant. The common factor $b_{j_1 k_1}$ of the elements of the first column may be taken out as a factor of the determinant. Treating the other columns similarly, we get

$$D = \sum_{j_1=1}^{n} \cdots \sum_{j_t=1}^{n} \Delta\, b_{j_1 k_1} \cdots b_{j_t k_t},$$

$$\Delta = \begin{vmatrix} a_{i_1 j_1} & \cdots & a_{i_1 j_t} \\ \cdots & \cdots & \cdots \\ a_{i_t j_1} & \cdots & a_{i_t j_t} \end{vmatrix}.$$

Unless $j_1, \ldots, j_t$ are distinct, $\Delta = 0$. Select $g_1, \ldots, g_t$ from $1, \ldots, n$ so that $g_1 < g_2 < \cdots < g_t$. If $j_1 = g_1, \ldots, j_t = g_t$, Δ is a t-rowed determinant α of A. Next, let $j_1, \ldots, j_t$ be an arrangement of $g_1, \ldots, g_t$ which is derived from $g_1, \ldots, g_t$ by l successive interchanges of two terms. Hence Δ may be derived from α by l successive interchanges of two columns, so that $\Delta = (-1)^l \alpha$. The sum of the $n!$ products $(-1)^l b_{j_1 k_1} \cdots b_{j_t k_t}$

corresponding to all such arrangements $j_1, \ldots, j_t$ is, by definition, the expansion of the determinant

$$\beta = \begin{vmatrix} b_{g_1 k_1} & \cdots & b_{g_1 k_t} \\ \cdots & \cdots & \cdots \\ b_{g_t k_1} & \cdots & b_{g_t k_t} \end{vmatrix},$$

which is a determinant of B. Hence $D = \sum \alpha \beta$, where the summation extends over all the selections $g_1, \ldots, g_t$ from $1, \ldots, n$ such that $g_1 < g_2 < \cdots < g_t$.

COROLLARY. *The rank of the product of two matrices cannot exceed the rank of either factor.*

For, if all t-rowed determinants of A (or of B) are zero, the same is true of all t-rowed determinants of AB.

THEOREM 6. *If A is any matrix with m rows and n columns and B is any non-singular, n-rowed, square matrix, then A and AB have the same rank. If C is any non-singular, m-rowed, square matrix, then A and CA have the same rank.*

For, if r is the rank of A and ρ is the rank of $P = AB$, the Corollary gives $\rho \leqq r$ and, when applied to $A = PB^{-1}$, gives $r \leqq \rho$, whence $r = \rho$. Next, if r' is the rank of CA, the Corollary gives $r' \leqq r$ and, when applied to $A = C^{-1} \cdot CA$, gives $r \leqq r'$, whence $r' = r$.

27. Bilinear forms. A polynomial in the $m + n$ variables $x_1, \ldots, x_m, y_1, \ldots, y_n$ is called a *bilinear form* if each of its terms is of the first degree in the x's and also of the first degree in the y's. An example with $m = 1$, $n = 2$, is $7x_1 y_1 - 5x_1 y_2$. The general bilinear form is

(23) $$\alpha = \sum_{i=1}^{m} \sum_{j=1}^{n} a_{ij} x_i y_j.$$

We may write

$$\alpha = \sum_{i=1}^{m} x_i l_i, \qquad l_i = \sum_{j=1}^{n} a_{ij} y_j \qquad (i = 1, \ldots, m).$$

The matrix $A = (a_{ij})$ of these m linear forms l_i is called the matrix of the bilinear form (23). The coefficient a_{ij} of $x_i y_j$ in α is therefore the element in the ith row and jth column of the matrix A of α.

By Theorem 1, the linear transformation (7) with matrix B replaces l_i by $\sum_k p_{ik} z_k$, where $(p_{ik}) = AB$. Hence that transformation replaces α by

$$\sum_{i=1}^{m} \sum_{k=1}^{n} p_{ik} x_i z_k,$$

whose matrix is AB.

If we do not alter the y's, but apply the transformation

(24) $\qquad x_i = \sum_{k=1}^{m} c_{ik} \xi_k \qquad (i = 1, \ldots, m)$

of matrix C to α in (23), we get

$$\sum_{i,k=1}^{m} \sum_{j=1}^{n} c_{ik} a_{ij} \xi_k y_j = \sum_{k=1}^{m} \sum_{j=1}^{n} d_{kj} \xi_k y_j,$$

where $d_{kj} = \sum_i c_{ik} a_{ij}$. Hence d_{kj} is the sum of the products of the elements of the kth row of matrix

$$C' = \begin{bmatrix} c_{11} & \cdots & c_{m1} \\ \cdots\cdots\cdots\cdots \\ c_{1m} & \cdots & c_{mm} \end{bmatrix}$$

by the corresponding elements of the jth column of $A = (a_{ij})$, so that the new bilinear form has the matrix $C'A$. Matrix C' is called the *transpose* of matrix C, being derived from C by employing the successive rows of C as the successive columns of C'.

THEOREM 7. *If in a bilinear form with the matrix A we subject the x's to a linear transformation (24) with the matrix C, and the y's to a linear transformation with the matrix B, we obtain a new bilinear form with the matrix $C'AB$, where C' is the transpose of C.*

We may write the bilinear form (23) as follows:

$$\alpha = \sum_{j=1}^{n} \lambda_j y_j, \qquad \lambda_j = \sum_{i=1}^{m} a_{ij} x_i \quad (j = 1, \ldots, n).$$

The matrix of the linear forms $\lambda_1, \ldots, \lambda_n$ is the transpose A' of matrix A. Transformation (24) with matrix C replaces $\lambda_1, \ldots, \lambda_n$ by linear functions of $\xi_1, \ldots, \xi_m$ the matrix of whose coefficients is $A'C$ (Theorem 1). Hence this transformation replaces α by a bilinear form with the matrix $(A'C)'$. By the result preceding the theorem, this matrix must be equal to $C'A$. Since the transpose of A' is A, this process proves

THEOREM 8. *The transpose of a product of matrices is equal to the product of their transposes taken in reverse order.*

EXERCISES

1. Prove Theorem 8 by direct multiplication.
2. Prove that $(A')^{-1} = (A^{-1})'$ by means of the transpose of $AA^{-1} = I$.
3. If $AA' = I$, then $A'A = I$.
4. An $m \times n$ matrix (a_{ij}) is of rank 0 or 1 if and only if there exist $m + n$ numbers $c_1, \ldots, c_m, d_1, \ldots, d_n$ such that $a_{ij} = c_i d_j$ ($i = 1, \ldots, m; j = 1, \ldots, n$).

28. Fields. Two forms (each linear, bilinear, or quadratic) may be equivalent under a transformation with complex coefficients, but not equivalent under one with real coefficients. Again, two forms may be equivalent under a transformation with real coefficients, but not under one with rational coefficients. Here, as elsewhere in modern algebra, it is necessary to specify carefully the nature of the constants employed. To do this briefly and clearly, as well as to give full generality to our theorems, we make use of the concept "field" (or domain of rationality).

A set of complex numbers $a + bi$ is called a *field* if the sum, difference, product, and quotient (the divisor not being zero) of any two equal or distinct numbers of the set are themselves numbers belonging to the set.[1]

The following are examples of sets of numbers each forming a field: all complex numbers; all real numbers; all rational numbers; all numbers of the form $a + br$ with a and b rational numbers and $r = i$ or $r = 3^{\frac{1}{2}}$.

But the set of all integers is not a field; nor do all positive real (or rational) numbers form a field.

29. Linear independence.[2] The quantities (numbers or functions) $l_1, \ldots, l_n$ are called *linearly dependent* with respect to a field F if there exist numbers $c_1, \ldots, c_n$, not all zero, of F such that

$$(25) \qquad c_1 l_1 + \cdots + c_n l_n = 0.$$

If no such numbers c_i exist, $l_1, \ldots, l_n$ are called *linearly independent* with respect to F.

It is convenient in proofs by induction on n to employ these terms also in case $n = 1$. Thus l_1 is linearly dependent or independent according as $l_1 = 0$ or $l_1 \neq 0$, respectively.

---

[1] Further fields satisfy the definition by postulates in the author's Algebras and Their Arithmetics, University of Chicago Press, Ch. XI. An example is the field of the residues of integers modulo p, a prime. For it, our theorems become important properties of congruences.

[2] We shall not presuppose the theory of systems of linear equations with coefficients in any field F, but deduce that theory in §§31, 32 from the present discussion. For this reason we avoid the shorter proof of Theorem 9 by means of that theory.

§ 29]   LINEAR INDEPENDENCE   55

THEOREM 9. *If $y_1, \ldots, y_n$ are linearly independent with respect to a field F, the n linear functions*

(26) $$l_i = \sum_{j=1}^{n} a_{ij} y_j \qquad (i = 1, \ldots, n),$$

with coefficients in F, are linearly independent or dependent with respect to F according as the determinant $\Delta = |a_{ij}|$ is not zero or is zero.

First let $\Delta \neq 0$ and assume, contrary to the theorem, that a relation (25) holds. Then

$$\sum_{i=1}^{n} c_i l_i = \sum_{i,j=1}^{n} c_i a_{ij} y_j = 0, \quad \sum_{i=1}^{n} a_{ij} c_i = 0 \quad (j = 1, \ldots, n).$$

Hence $\Delta c_i = 0$, and every $c_i = 0$, contrary to hypothesis.

Conversely, if $l_1, \ldots, l_n$ are linearly independent, then $\Delta \neq 0$. We shall give a proof by induction on n, noting that it is evidently true when $n = 1$. If $l_n = 0$, $1 \cdot l_n = 0$ would be a relation (25). Hence after permuting the y's, we may assume that $a_{nn} \neq 0$. Since our converse will follow if proved for $l_1, \ldots, l_{n-1}, a_{nn}^{-1} l_n$, we may assume that $a_{nn} = 1$. Then

$$z_i = l_i - a_{in} l_n = \sum_{j=1}^{n-1} (a_{ij} - a_{in} a_{nj}) y_j \quad (i = 1, \ldots, n-1)$$

are linearly independent linear functions of $y_1, \ldots, y_{n-1}$, since

$$\sum_{i=1}^{n-1} c_i z_i = \sum_{i=1}^{n-1} c_i (l_i - a_{in} l_n) = 0$$

implies that each $c_i = 0$ in view of the linear independence of $l_1, \ldots, l_n$. Hence by the hypothesis for the induction, the determinant

$$D = |a_{ij} - a_{in} a_{nj}| \qquad (i, j = 1, \ldots, n-1)$$

is not zero. If in Δ we add to the first, second, ..., $(n-1)$th columns the products of the elements of the last column by $-a_{n1}$, $-a_{n2}$, ..., $-a_{nn-1}$, respectively, we get a determinant whose last row is $0, 0, \ldots, 0, 1$ and having D as the cofactor of 1. Hence $\Delta = D \neq 0$.

THEOREM 10. *If $y_1, \ldots, y_n$ are linearly independent with respect to F, and $l_1, \ldots, l_m$ are any linear forms in $y_1, \ldots, y_n$ with coefficients in F, we can select certain of the l's, say*

$$\eta_i = l_{k_i} \qquad (i = 1, \ldots, r),$$

which are linearly independent with respect to F and are such that the $m - r$ remaining l's are expressible linearly in terms of $\eta_1, \ldots, \eta_r$ with coefficients in F.

As η_1 we may take any l_{k_1} which is not zero identically in the y's. If every l_j is a product of η_1 by a number of F, the theorem holds with $r = 1$. In the contrary case, take as η_2 any l_{k_2} which is not such a product, whence η_1 and η_2 are linearly independent. If every l_j is a linear combination of η_1 and η_2 with coefficients in F, the theorem holds with $r = 2$. In the contrary case, take as η_3 any l_{k_3} which is not such a combination, etc.

The value of r is given by Theorem 13.

THEOREM 11. *If $r < n$ in Theorem 10, there exist linear forms $\eta_{r+1}, \ldots, \eta_n$ in $y_1, \ldots, y_n$ with coefficients in F such that $\eta_1, \ldots, \eta_n$ are linearly independent with respect to F.*

First, $y_1, \ldots, y_n$ are not all linear functions of $\eta_1, \ldots, \eta_r$ with coefficients in F. For, if so, we have in particular

$$(27) \qquad y_i = \sum_{k=1}^{r} t_{ik} \eta_k \qquad (i = 1, \ldots, r).$$

Hence by Theorem 9, $|t_{ik}| \neq 0$. We may therefore solve (27) and obtain η_k as a linear function of $y_1, \ldots, y_r$ with coefficients in F. By hypothesis, y_i $(i > r)$ is a linear function of the η's and hence of $y_1, \ldots, y_r$ with coefficients in F. This contradicts the linear independence of $y_1, \ldots, y_n$ with respect to F.

Hence not all linear functions of $y_1, \ldots, y_n$ with coefficients in F are linear functions of $\eta_1, \ldots, \eta_r$ with coefficients in F. Thus there exists a linear function η_{r+1} of the y's such that $\eta_1, \ldots, \eta_r$, η_{r+1} are linearly independent with respect to F. If $r + 1 = n$, the theorem is proved. If $r + 1 < n$, we repeat the argument in the first part with r replaced by $r + 1$ and conclude that $y_1, \ldots, y_n$ are not all linear functions of $\eta_1, \ldots, \eta_{r+1}$ with coefficients in F. Thus there exists a linear function η_{r+2} of the y's such that $\eta_1, \ldots,$ η_{r+2} are linearly independent with respect to F. This proves the theorem if $r + 2 = n$. If $r + 2 < n$, we repeat the argument.

30. Equivalence of two matrices or two bilinear forms. Two bilinear forms α and α_1 in $x_1, \ldots, x_m, y_1, \ldots, y_n$, with coefficients in F, are called *equivalent in F* if and only if α becomes α_1 when we subject the x's in α to a linear transformation with non-singular matrix C and the y's to a linear transformation with non-singular matrix B, the elements of C and B being in F. Then also the matrices A and A_1 of α and α_1 are called equivalent in F. By Theorem 7, $A_1 = C'AB$. In other words, two $m \times n$ matrices A and A_1 with elements in F are called *equivalent in F* if and only if there exist two non-singular square matrices[1] D and B with elements in F and having m and n rows, respectively, such that $DAB = A_1$.

THEOREM 12. *Two matrices (or bilinear forms) in F are equivalent in F if and only if they have the same rank.*

If the matrices (or bilinear forms) are equivalent, they have the same rank by Theorem 6.

Conversely, if α and α_1 are any bilinear forms in F with the same rank, they are equivalent in F. This is evident if the rank is zero, whence $\alpha \equiv 0$, $\alpha_1 \equiv 0$. Consider therefore a bilinear form α,

---

[1] We have written D for the earlier C'. Conversely, given D, we get $C = D'$.

$$\alpha = \sum_{i=1}^{m} x_i l_i, \qquad l_i = \sum_{j=1}^{n} a_{ij} y_j \qquad (i = 1, \ldots, m),$$

in which the a_{ij} are numbers of F not all zero. By Theorem 10, we may select certain of the l's, say $\eta_i = l_{k_i}$ $(i = 1, \ldots, r)$, which are linearly independent with respect to F and such that the $m - r$ remaining l's are linear functions of $\eta_1, \ldots, \eta_r$ with coefficients in F. Since $k_1, \ldots, k_r$ are distinct, we may select $m - r$ further k's so that $k_1, \ldots, k_m$ form a permutation of $1, \ldots, m$. Also for $i > r$, denote l_{k_i} by η_i. Introduce the new variables

$$X_i = x_{k_i} \qquad (i = 1, \ldots, m).$$

Then

$$\alpha = \sum_{i=1}^{m} x_{k_i} l_{k_i} = \sum_{i=1}^{m} X_i \eta_i, \quad \eta_i = \sum_{j=1}^{r} d_{ij} \eta_j \quad (i = r+1, \ldots, m),$$

where the d_{ij} belong to F. Hence

$$\alpha = \sum_{j=1}^{r} X_j \eta_j + \sum_{i=r+1}^{m} X_i \sum_{j=1}^{r} d_{ij} \eta_j = \sum_{j=1}^{r} \xi_j \eta_j,$$

if we write

$$\xi_j = X_j + \sum_{i=r+1}^{m} d_{ij} X_i \qquad (j = 1, \ldots, r).$$

The latter and $\xi_i = X_i (i = r + 1, \ldots, m)$ give m linearly independent linear functions $\xi_1, \ldots, \xi_m$ of $X_1, \ldots, X_m$ with coefficients in F. Hence the ξ's are linearly independent linear functions of $x_1, \ldots, x_m$ with coefficients in F. The determinant of the matrix M of the coefficients of $x_1, \ldots, x_m$ in the ξ's is not zero by Theorem 9. If $r < n$, there exist, by Theorem 11, linear functions

§ 30]   EQUIVALENT MATRICES   59

$\zeta_{r+1}, \ldots, \zeta_n$ of $y_1, \ldots, y_n$ with coefficients in F such that $\eta_1, \ldots, \eta_r, \zeta_{r+1}, \ldots, \zeta_n$ are linearly independent with respect to F. If $r = n$, we suppress the ζ_i and employ $\eta_1, \ldots, \eta_n$. In either case, the determinant of the matrix N of the coefficients of $y_1, \ldots, y_n$ in these n linear functions is not zero by Theorem 9.

We have now proved that α can be reduced to the canonical form

(28) $$\sum_{j=1}^{r} \xi_j \eta_j$$

by a linear transformation expressing the x's as linear functions of the ξ's with the matrix M^{-1} and a linear transformation expressing the y's as linear functions of the η's and ζ's with the matrix N^{-1}. The matrix of (28) is evidently of rank r. By the first part of our theorem, also α is of rank r.

Since α and α_1 are both equivalent to (28), they are equivalent to each other.

We have also proved the first part of

THEOREM 13. *The r of Theorem 10 is the rank of the matrix of the coefficients of $l_1, \ldots, l_m$. Moreover, the rank of the matrix T of the coefficients of $\eta_1, \ldots, \eta_r$ is r.*

Let ρ be the rank of T. Apply the first part of the theorem to $\eta_1, \ldots, \eta_r$ in place of $l_1, \ldots, l_m$. Hence if $r > \rho$, $r - \rho$ of the η's are expressible linearly in terms of ρ of them with coefficients in F, contrary to the linear independence of $\eta_1, \ldots, \eta_r$ with respect to F.

COROLLARY 1. *If $m > n$, any m linear forms in n independent variables with coefficients in F are linearly dependent with respect to F.*

For, the rank r is evidently $\leq n$ and hence $< m$. Then by Theorem 10, $m - r$ of the forms are expressible in terms of r of them, so that the m forms are linearly dependent.

Note that, conversely, if the m forms are linearly dependent, then the rank r is less than m. Hence we have

COROLLARY 2. *Any m linear forms in n independent variables with coefficients in F are linearly dependent with respect to F if and only if every m-rowed determinant of the matrix of the coefficients is zero.*

EXERCISES

1. Find linear transformations which reduce

$$2x_1 y_2 - 3x_1 y_3 + 4x_2 y_1 - 2x_2 y_3 - 3x_3 y_1 + x_3 y_2 \text{ to } \xi_1 \eta_1 + \xi_2 \eta_2.$$

2. $5x + 3y - 6$, $8x + 15z$, $6x + 2y + 5z - 4$ are linearly dependent.

3. When F is the field of all real numbers, restate Theorem 12 without using explicitly the notion of a field.

4. Any matrix of rank r is equivalent to one whose element in the ith row and ith column is 1 when $i \leq r$, while all further elements are zero.

5. Any matrix of rank r can be expressed as the sum of r matrices of rank unity. Hence any bilinear form of rank r can be written as the sum of r products each of a linear form in $x_1, \ldots, x_m$ by a linear form in $y_1, \ldots, y_n$.

6. When a bilinear form is the sum of r products each of a linear form in $x_1, \ldots, x_m$ by a linear form in $y_1, \ldots, y_n$, the matrix of the form is of rank r if and only if the r linear forms in the x's are linearly independent and the r linear forms in the y's are linearly independent.

7. If r is the rank of an n-rowed square matrix and if R is the rank of the matrix composed of the first t rows of the former, then $R \geq r + t - n$.

8. If r_1 and r_2 are the ranks of two n-rowed square matrices, and if R is the rank of their product, show that $R \geq r_1 + r_2 - n$. Hint: First prove this when the left factor is of the form in Ex. 4.

31. Homogeneous linear equations. Consider a system of equations $l_1 = 0, \ldots, l_m = 0$ in the unknowns $y_1, \ldots, y_n$ with coefficients in a field F and of rank r. After rearranging the equations, we may assume by Theorem 13 that $l_1, \ldots, l_r$ are linearly independent and have a matrix M of rank r, while $l_{r+1}, \ldots, l_m$ are linear combinations of $l_1, \ldots, l_r$ with coefficients in F. Some

§ 31] HOMOGENEOUS LINEAR EQUATIONS 61

r-rowed determinant of M is not zero. After relabeling the y's, we may therefore assume that the determinant D of the coefficients of $y_1, \ldots, y_r$ in $l_1 = 0, \ldots, l_r = 0$ is not zero. Transpose to the second members the terms involving $y_j (j > r)$ and apply the ordinary method of solution by determinants. We obtain

(29) $$y_i = \sum_{j=r+1}^{n} c_{ij} y_j \quad (i = 1, \ldots, r),$$

where the c_{ij} are in F. When these values of $y_1, \ldots, y_r$ are inserted in $l_1, \ldots, l_r$, we therefore obtain expressions in $y_{r+1}, \ldots, y_n$ which are zero for all values of the latter. The same follows for $l_{r+1}, \ldots, l_m$ since they are linear combinations of $l_1, \ldots, l_r$.

THEOREM 14. *Given m homogeneous linear equations in n unknowns whose coefficients belong to any field F and have a matrix of rank r, we may select r of the equations so that their matrix has a non-vanishing r-rowed determinant. These r equations determine uniquely r of the unknowns as homogeneous linear functions, with coefficients in F, of the remaining n − r unknowns. For all values of the latter, the expressions for the r unknowns satisfy the given m equations.*

COROLLARY. There exist solutions not all zero if and only if $r < n$ and certainly if $m < n$. In particular, n homogeneous linear equations in n unknowns have solutions not all zero if and only if the determinant of the coefficients is zero.

If the homogeneous linear equations $l_1 = 0, \ldots, l_m = 0$ have the two solutions $(u_1, \ldots, u_n)$ and $(v_1, \ldots, v_n)$, they evidently have the solution $(gu_1 + hv_1, \ldots, gu_n + hv_n)$, in which g and h are any numbers. We call the two solutions *linearly dependent* if there exist constants g and h, not both zero, such that $gu_i + hv_i = 0$ $(i = 1, \ldots, n)$; but *linearly independent* if no such constants g and h exist. Analogous definitions apply to more than two solutions.

For example, the single equation $2y_1 + 3y_2 - 5y_3 = 0$ has the linearly independent solutions

$$(u_1, u_2, u_3) = (-3/2, 1, 0), \qquad (v_1, v_2, v_3) = (5/2, 0, 1),$$

while every solution (y_1, y_2, y_3) is linearly dependent on them:

$$y_1 = y_2 u_1 + y_3 v_1, \qquad y_2 = y_2 u_2 + y_3 v_2, \qquad y_3 = y_2 u_3 + y_3 v_3.$$

This example illustrates the following

THEOREM 15. *Any system of m homogeneous linear equations in $y_1, \ldots, y_n$ of rank $r < n$ has $n - r$ linearly independent solutions, while every solution is linearly dependent on them.*

With the notations (29), we may take as the $n - r$ solutions

$$(30) \quad (c_{1j}, \ldots, c_{rj}, \; 0, \ldots, 0, 1, 0, \ldots, 0) \quad (j = r+1, \ldots, n),$$

where 1 is in the jth place. Multiply the numbers in (30) by y_j, sum as to j, and apply (29); we get the solution

$$\left(\sum y_j c_{1j} = y_1, \ldots, \sum y_j c_{rj} = y_r, \; y_{r+1}, \ldots, y_n \right),$$

in which the summations extend from $j = r + 1$ to $j = n$. Since these numbers are all zero if and only if $y_{r+1}, \ldots, y_n$ are all zero, the solutions (30) are linearly independent.

32. Non-homogeneous linear equations. Consider a system of m equations $l_1 = k_1, \ldots, l_m = k_m$, where the k's are constants and $l_1, \ldots, l_m$ are linear forms in $y_1, \ldots, y_n$ whose matrix A is of rank r. After rearranging the equations, we may assume that $l_1, \ldots, l_r$ are linearly independent, while

$$(31) \qquad l_i = \sum_{j=1}^{r} c_{ij} l_j \qquad (i = r + 1, \ldots, m).$$

§ 32] NON-HOMOGENEOUS LINEAR EQUATIONS

Hence our m equations $l_i = k_i$ are inconsistent unless

(32) $\qquad k_i = \sum_{j=1}^{r} c_{ij} k_j \qquad (i = r+1, \ldots, m).$

Write $\lambda_i \equiv l_i - k_i y_{n+1}$, where $y_{n+1} = 1$. Since $l_1, \ldots, l_r$ are linearly independent, the same is true of $\lambda_1, \ldots, \lambda_r$. Next,

$$\lambda_i = \sum_{j=1}^{r} c_{ij} \lambda_j \qquad (i = r+1, \ldots, m)$$

if and only if conditions (32) hold, and then the matrix of $\lambda_1, \ldots, \lambda_m$ is also of rank r by Theorem 13.

Theorem 16. *Let r be the rank of the matrix A of the homogeneous linear functions $l_1, \ldots, l_m$ of $y_1, \ldots, y_n$. Let B be the augmented matrix derived from A by annexing a column having the elements $k_1, \ldots, k_m$, so that the rank ρ of B is $\geq r$. The equations $l_1 = k_1$, $\ldots, l_m = k_m$ are inconsistent if $\rho > r$. If $\rho = r$, certain r of the equations determine uniquely r of the unknowns as linear functions of the remaining $n - r$ unknowns; for all values of the latter, the expressions for these r unknowns satisfy also the remaining $m - r$ equations.*

CHAPTER IV

QUADRATIC AND HERMITIAN FORMS, SYMMETRIC AND HERMITIAN BILINEAR FORMS

Our discussion of these four types of forms requires but little more space and effort than are necessary for a treatment of quadratic forms alone. There has been a steady increase in the importance of the forms named after Hermite. Our results on quadratic forms suffice both for their metric and their projective classification.

33. Symmetric bilinear forms and quadratic forms. The form

$$(1) \qquad \alpha = \sum_{i,j=1}^{n} a_{ij} x_i y_j$$

is called a *symmetric* bilinear form if it remains unaltered by the interchange of x_i and y_i for $i = 1, \ldots, n$, and hence if

$$(2) \qquad a_{ji} = a_{ij} \qquad (i, j = 1, \ldots, n).$$

The matrix $A = (a_{ij})$ of α then remains unaltered by the interchange of rows and corresponding columns, and is called a *symmetric* matrix. In other words, A is symmetric if and only if it is equal to its transpose A'.

By Theorem 7 of Ch. III, the transformation

$$(3) \qquad x_i = \sum_{j=1}^{n} b_{ij} X_j \qquad (i = 1, \ldots, n),$$

with the matrix $B = (b_{ij})$, and the *cogredient* transformation

$$(4) \qquad y_i = \sum_{j=1}^{n} b_{ij} Y_j \qquad (i = 1, \ldots, n),$$

§ 33] SYMMETRIC BILINEAR FORMS 65

with the same matrix B, replace (1) by a bilinear form α_1, whose matrix is $A_1 = B'AB$. If $A' = A$, then $A'_1 = B'A'B = A_1$ by Theorem 8 of Ch. III, whence also matrix A_1 is symmetric.

Any two n-rowed square matrices A and A_1 with elements in a field F are called *congruent* in F if there exists a non-singular n-rowed square matrix B with elements in F such that $A_1 = B'AB$. By Theorem 6 of Ch. III, congruent matrices have the same rank.

Recalling the definition of equivalence in §30, we have

THEOREM 1. *Two symmetric bilinear forms with coefficients in a field F are equivalent under non-singular cogredient transformations (3) and (4) with coefficients in F if and only if their matrices are congruent in F.*

If we identify y_i with x_i for $i = 1, \ldots, n$, a symmetric bilinear form (1) becomes a quadratic form

$$(5) \qquad q = \sum_{i,\,j=1}^{n} a_{ij} x_i x_j, \qquad a_{ji} = a_{ij},$$

having the same symmetric matrix $A = A'$. For $n = 2$,

$$q = a_{11} x_1^2 + 2a_{12} x_1 x_2 + a_{22} x_2^2.$$

The rank and determinant of A are called the rank and determinant of q.

If also in (4) we identify y_i with x_i, and Y_j and X_j, so that (4) becomes (3), we conclude that *transformation (3) with the matrix B replaces a quadratic form (5) with the symmetric matrix A by a quadratic form with the symmetric matrix $B'AB$.*

Hence we have

THEOREM 2. *Two quadratic forms with coefficients in a field F are equivalent under non-singular transformation with coefficients in F if and only if their matrices are congruent in F.*

A comparison of Theorems 1 and 2 yields

COROLLARY 1. Two symmetric bilinear forms with matrices A and A_1 are equivalent under cogredient transformations if and only if the quadratic forms with matrices A and A_1 are equivalent, where the forms and equivalence relate to the same field F.

Hence these two problems of equivalence are not essentially different and each may be treated by a study of the congruence of symmetric matrices.

34. Hermitian forms and Hermitian bilinear forms. Let S be any field containing a number ν which is not the square of any number of S. Then all the numbers $x = a + b\nu^{\frac{1}{2}}$, in which a and b belong to S, form a larger field F. Call $\bar{x} = a - b\nu^{\frac{1}{2}}$ the *conjugate* of x.

An important example is that in which S is the field of all real numbers, and $\nu = -1$. Then F is the field of all complex numbers, and x and $\bar{x}$ are conjugate imaginaries.

The bilinear form, with coefficients in F,

$$(6) \qquad \sum_{i,\,j=1}^{n} a_{ij} x_i y_j, \qquad a_{ji} = \bar{a}_{ij},$$

is called an *Hermitian bilinear form in F*.

If we replace each element of a matrix A by its conjugate, we obtain a matrix denoted by $\bar{A}$ and called the *conjugate* of A. Hence the matrix A of (6) has the property

$$(7) \qquad A' = \bar{A},$$

and is called an *Hermitian* matrix. It is seen at once that

$$(8) \qquad \bar{\bar{M}} = M, \qquad \overline{MN} = \bar{M}\,\bar{N}, \qquad (\bar{M})' = \bar{M}'.$$

By Theorem 7 of Ch. III, transformation (3) with the matrix B and the *conjugate* transformation

(9) $$y_i = \sum_{j=1}^{n} \bar{b}_{ij} Y_j, \quad (i = 1, \ldots, n),$$

with the conjugate matrix $\bar{B}$, replace (6) by a form whose matrix is $A_1 = B'A\bar{B}$. Then (7) implies

$$A_1' = \bar{B}'A'B = \bar{A}_1.$$

THEOREM 3. *A linear transformation with matrix B and the conjugate transformation replace an Hermitian bilinear form with the Hermitian matrix A by an Hermitian bilinear form with the Hermitian matrix $B'A\bar{B}$.*

Any two n-rowed square matrices A and A_1 with elements in a field F shall be called *conjunctive* in F if there exists a non-singular n-rowed square matrix B with elements in F such that $A_1 = B'A\bar{B}$. By Theorem 6 of Ch. III, conjunctive matrices have the same rank. Theorem 3 implies

THEOREM 4. *Two Hermitian bilinear forms with coefficients in the field F are equivalent under a non-singular transformation and its conjugate with coefficients in F if and only if their matrices are conjunctive in F.*

If in (6) we identify y_j with $\bar{x}_j$, we get the *Hermitian form*

(10) $$\sum_{i,j=1}^{n} a_{ij} x_i \bar{x}_j, \quad a_{ji} = \bar{a}_{ij}.$$

This form is equal to its conjugate. If we also identify Y_j with $\bar{X}_j$, we see that (9) become equations which follow from (3) by taking conjugates. Hence Theorem 3 implies

THEOREM 5. *A linear transformation with the matrix B replaces an Hermitian form with the Hermitian matrix A by an Hermitian form with the Hermitian matrix $B'A\bar{B}$.*

This implies the following analog of Theorem 4:

68   QUADRATIC AND HERMITIAN FORMS   [Ch. IV

THEOREM 6. *Two Hermitian forms with coefficients in the field F are equivalent under a non-singular transformation with coefficients in F if and only if their matrices are conjunctive in F.*

A comparison of Theorems 4 and 6 yields

COROLLARY 2. Two Hermitian bilinear forms with matrices A and A_1 are equivalent under a non-singular transformation and its conjugate if and only if the Hermitian forms with matrices A and A_1 are equivalent, where the forms and equivalence relate to the same field F.

Hence these two problems of equivalence are not essentially different. From this and the corresponding result in Corollary 1, we see that the conclusions in our further study of two quadratic or two Hermitian forms imply corresponding conclusions concerning two symmetric or two Hermitian bilinear forms.

Not only have the latter two problems been reduced to the former two, but the former two may be treated together by the simple device next explained.

35. Rational reduction of quadratic and Hermitian forms. To each element a of the field F we make correspond an element $\tilde{a}$ of F such that

(11) $\quad\quad \tilde{\tilde{a}} = a, \quad \widetilde{ag} = \tilde{a}\tilde{g}, \quad \widetilde{a \pm g} = \tilde{a} \pm \tilde{g}.$

Hence to a matrix A with elements a_{ij} in F corresponds a matrix $\tilde{A}$ whose element in the ith row and jth column is $\tilde{a}_{ij}$. Hence, as in (8), we have

(12) $\quad\quad \tilde{\tilde{A}} = A, \quad \widetilde{AQ} = \tilde{A}\tilde{Q}, \quad (\tilde{A})' = \tilde{A}'.$

If F is obtained as in §34 from a subfield S by adjoining $\nu^{\frac{1}{2}}$, and if we take $\tilde{a}$ to be $\bar{a}$ for every a in F, we see that relations (11) hold. They evidently hold also if we take $\tilde{a}$ to be a itself. These

§ 35] RATIONAL REDUCTION 69

two cases are the ones needed in Hermitian and quadratic forms respectively, as we shall next explain in detail.
Consider the form, with coefficients in F,

(13) $$\sum_{i,\,j=1}^{n} a_{ij}\, x_i\, \tilde{x}_j, \qquad a_{ji} = \tilde{a}_{ij},$$

whose matrix A therefore has the property

(14) $$A' = \tilde{A}.$$

If $\tilde{a} = \bar{a}$ for every a in F, and if $\tilde{x}_j = \bar{x}_j$, (13) becomes the Hermitian form (10) whose matrix A is Hermitian and hence has the property $A' = \bar{A}$. But if $\tilde{a} = a$ for every a in F, and if $\tilde{x}_j = x_j$, (13) becomes the quadratic form (5) whose matrix A is symmetric and hence has the property $A' = A$. Hence the form (13) furnishes a generalization of Hermitian and quadratic forms, which will be seen to effect a unification of the two theories of those forms.

THEOREM 7. *By a non-singular transformation*[1] *with coefficients in a field F, any form* (13) *of rank r with coefficients in F can be reduced to one of type*

(15) $$g_1 x_1 \tilde{x}_1 + \cdots + g_r x_r \tilde{x}_r \qquad (\text{each } g_i \neq 0,\ \tilde{g}_i = g_i).$$

We shall first transform (13) into a like form $l\, X_1\, \tilde{X}_1 + \cdots$ having $l \neq 0$. If $a_{11} = 0$ and $a_{kk} \neq 0$, the transformation

$$x_1 = X_k, \quad x_k = X_1, \quad x_i = X_i \quad (i \neq 1, k)$$

replaces (13) by $a_{kk} X_1 \tilde{X}_1 + \ldots$. If every $a_{ii} = 0$ and $a_{12} \neq 0$, the transformation

$$x_1 = X_1, \quad x_2 = gX_1 + X_2, \quad x_i = X_i \quad (i > 2)$$

---
[1] Transformation (3), of matrix B, together with the induced transformation $\tilde{x}_i = \sum \tilde{b}_{ij} \tilde{X}_j$. The matrix of the resulting form is $B'A\tilde{B}$ and has the same rank as A by Theorem 6 of Ch. III.

replaces (13) by $lX_1 \tilde{X}_1 + \ldots$, where $l = a_{12}\, \tilde{g} + a_{21}\, g$. Taking $g = 1/a_{21}$, we have $l = 2$.

It remains to consider a form α of type (13) having $a_{11} \neq 0$. Then

$$\alpha = \frac{1}{a_{11}}(a_{11} x_1 + \sum_{i=2}^{n} a_{i1}\, x_i)(a_{11}\, \tilde{x}_1 + \sum_{j=2}^{n} a_{1j}\, \tilde{x}_j)$$
$$+ \phi(x_2, \ldots, x_n,\ \tilde{x}_2, \ldots, \tilde{x}_n).$$

Denoting the quantity in the first parenthesis by $a_{11} X_1$, we see that that in the second parenthesis is equal to $a_{11} \tilde{X}_1$, and hence that the transformation

$$x_1 = X_1 - \frac{1}{a_{11}}\sum_{i=2}^{n} a_{i1}\, X_i, \qquad x_k = X_k \quad (k > 1)$$

replaces α by

$$a_{11}\, X_1\, \tilde{X}_1 + \phi(X_2, \ldots, X_n,\ \tilde{X}_2, \ldots, \tilde{X}_n).$$

Since the matrices of α and $a_{11} X_1 \tilde{X}_1$ have the property (14), the matrix of ϕ has that property. We may therefore proceed with ϕ as we did with (13). Ultimately we obtain (15). The number of its terms is evidently its rank and hence is the rank of (13) by the preceding footnote.

COROLLARY 3. *Within any field any quadratic form of rank* r *is equivalent to*

(16) $\qquad g_1 x_1^2 + \cdots + g_r x_r^2 \qquad$ (*each* $g_i \neq 0$).

36. Canonical forms. When F is the field of all complex numbers, (16) is reduced to $\xi_1^2 + \cdots + \xi_r^2$ by the transformation

$$x_i = g_i^{-\frac{1}{2}}\, \xi_i \qquad (i = 1, \ldots, r).$$

THEOREM 8. *Within the field of all complex numbers, every quadratic form of rank* r *can be reduced to* $\xi_1^2 + \cdots + \xi_r^2$ *by a non-*

singular linear transformation. Hence two quadratic forms are equivalent if and only if they are of the same rank.

From this theorem and Corollary 1 we have

COROLLARY 4. *Two symmetric bilinear forms are equivalent under non-singular cogredient transformations with complex coefficients if and only if they have the same rank* r. *Each can be reduced to* $\xi_1 \eta_1 + \cdots + \xi_r \eta_r$.

THEOREM 9. *Let F be the field of all real numbers or the field of all complex numbers. In the respective cases, take $\tilde{a} = a$ or $\bar{a}$ for every a in F. By a non-singular transformation with coefficients in F, any form* (13) *of rank r with coefficients in F can be reduced to*

(17) $\quad y_1 \tilde{y}_1 + \cdots + y_p \tilde{y}_p - y_{p+1} \tilde{y}_{p+1} - \cdots - y_r \tilde{y}_r,$

where the **index** *p is uniquely determined. Hence two forms* (13) *are equivalent in F if and only if they have the same rank r and the same index p.*

Since $\tilde{g}_i = g_i$ in (15), each g_i is real. After applying to (15) the transformation which permutes certain of the x's, we may assume that $g_1, \ldots, g_p$ are positive and $g_j (j > p)$ is negative. Write $g_j = -d_j (j > p)$. Then the transformation

$$x_i = g_i^{-\frac{1}{2}} y_i \quad (i = 1, \ldots, p),$$
$$x_j = d_j^{-\frac{1}{2}} y_j \quad (j = p+1, \ldots, r),$$

with real coefficients, replaces (15) by (17). Hence any form (13) of rank r with coefficients in F can be reduced by a non-singular transformation with coefficients in F to one of type (17).

Suppose that it can be reduced also to

(18) $\quad z_1 \tilde{z}_1 + \cdots + z_q \tilde{z}_q - z_{q+1} \tilde{z}_{q+1} - \cdots - z_r \tilde{z}_r.$

We shall prove that $p = q$. Assume that $p > q$. By hypothesis, (17) and (18) are each equal to the given form (13) when the y's and z's are certain linear functions of $x_1, \ldots, x_n$ with coefficients in F. Hence (17) and (18) are equal for all values of the x's. The system of $q + n - p < n$ homogeneous linear equations

(19) $\quad z_1 = 0, \ldots, z_q = 0, \quad y_{p+1} = 0, \ldots, y_n = 0$

in the n unknowns $x_1, \ldots, x_n$ has solutions $x_1 = X_1, \ldots, x_n = X_n$ in F, not all zero, by the Corollary to Theorem 14 of Ch. III. Let Y_i and Z_i be the corresponding values of y_i and z_i respectively. Equating the resulting values of (17) and (18), we get

$$Y_1 \tilde{Y}_1 + \cdots + Y_p \tilde{Y}_p = - Z_{q+1} \tilde{Z}_{q+1} - \cdots - Z_r \tilde{Z}_r.$$

The terms on the left are zero or positive, while those on the right are zero or negative. Hence every term is zero. In particular, $Y_1, \ldots, Y_p$ are zero. In view also of (19), we see that the system of equations $y_1 = 0, \ldots, y_n = 0$ has the solutions $X_1, \ldots, X_n$, not all zero. Hence the determinant of the coefficient of the x's in $y_1, \ldots, y_n$ is zero by the Corollary cited. In other words, the transformation which reduced (13) to (17) is singular, contrary to hypothesis.

Separating the two cases of Theorem 9, we have

COROLLARY 5. *Any real quadratic form of rank r can be reduced by a non-singular real transformation to*

(20) $\quad y_1^2 + \cdots + y_p^2 - y_{p+1}^2 - \cdots - y_r^2,$

where the index p is uniquely[1] *determined. Hence two real quadratic forms are equivalent under non-singular real transformation if and only if they have the same rank and the same index.*

For $r = 2$, the types are $y_1^2 + y_2^2$, $y_1^2 - y_2^2$, $- y_1^2 - y_2^2$.

---

[1] Discovered independently by Jacobi, Jour. für Math., 53, 1857, 275, 281; Werke, 4, 1884, 593, and Sylvester (who called the uniqueness of p the law of inertia of quadratic forms), Phil. Mag., (4), 4, 1852, 140; Phil. Trans., London, 143, 1853, 481; Jour. für Math., 100, 1887, 465; Coll. Math. Papers, I, 1904, 376-7, 511; IV, 523.

COROLLARY 6. Within the field of complex numbers, any Hermitian form of rank r can be reduced to

(21) $\quad y_1 \bar{y}_1 + \cdots + y_p \bar{y}_p - y_{p+1} \bar{y}_{p+1} - \cdots - y_r \bar{y}_r,$

where the index p is uniquely determined. Hence two Hermitian forms are equivalent if and only if they have the same rank r and the same index p.

The rank r of a quadratic or Hermitian form f is the maximum order of a non-vanishing determinant of the matrix of f and hence may be found from f without resort to the canonical form. A corresponding direct method of finding the index p is given in Theorem 22.

Corollaries 1, 2, 5, 6 imply

COROLLARY 7. There is a unique determination of p in the canonical form

$$x_1 y_1 + \cdots + x_p y_p - x_{p+1} y_{p+1} - \cdots - x_r y_r$$

to which a real symmetric bilinear form can be reduced by non-singular real cogredient transformations, or to which a Hermitian bilinear form with complex coefficients can be reduced by a non-singular transformation on the x's and the conjugate transformation on the y's.

A real quadratic form is called *positive* if its index and rank are equal, and *negative* if its index is zero. Both are called *definite* forms. A positive form of rank r is equivalent under real transformation to $y_1^2 + \cdots + y_r^2$, while a negative one is equivalent to $- y_1^2 - \cdots - y_r^2$.

Positive and negative Hermitian forms are defined similarly. They are equivalent to (21) with $p = r$ and $p = 0$, respectively.

37. Reduction of a real quadratic form by real orthogonal transformation. We shall prove simultaneously the two companion theorems:

THEOREM 10. *If f and h are real n-ary quadratic forms and if h is positive and of rank n, the pair f, h can be reduced by a **real** non-singular linear transformation to*

$$f = \lambda_1 \xi_1^2 + \cdots + \lambda_n \xi_n^2, \qquad h = \xi_1^2 + \cdots + \xi_n^2,$$

where $\lambda_1, \ldots, \lambda_n$ are the roots of $|f - \lambda h| = 0$ and are all real.

THEOREM 11. *If f and h are n-ary Hermitian forms and if h is positive and of rank n, the pair f, h can be reduced by a non-singular linear transformation to*

$$f = \lambda_1 \xi_1 \bar{\xi}_1 + \cdots + \lambda_n \xi_n \bar{\xi}_n, \qquad h = \xi_1 \bar{\xi}_1 + \cdots + \xi_n \bar{\xi}_n,$$

where $\lambda_1, \ldots, \lambda_n$ are the roots of $|f - \lambda h| = 0$ and are all real.

Let $A = (a_{ij})$ and $B = (b_{ij})$ be the matrices of f and h. Let λ_1 be any root of $|A - \lambda B| = 0$. Then the system of n linear equations

$$\sum_{j=1}^{n} a_{ij} \tilde{x}_j = \lambda_1 \sum_{j=1}^{n} b_{ij} \tilde{x}_j \qquad (i = 1, \ldots, n)$$

become after transposition of terms a system of n homogeneous linear equations in the $\tilde{x}_j$ whose determinant is $|A - \lambda_1 B| = 0$, and hence have solutions $\tilde{x}_1, \ldots, \tilde{x}_n$ not all zero. Then

$$f = \sum_{i,j} a_{ij} x_i \tilde{x}_j = \sum_i x_i \lambda_1 \sum_j b_{ij} \tilde{x}_j = \lambda_1 h, \quad h = \sum_{i,j} b_{ij} x_i \tilde{x}_j.$$

First, let f and h be Hermitian. Then $h \neq 0$ by the assumption that h is positive and of rank n. Since f and h are real, also λ_1 is real.

§37] REDUCTION BY ORTHOGONAL TRANSFORMATION 75

Next, let h be a real positive quadratic form. Consider the Hermitian form $H = \sum b_{st} x_s \bar{x}_t$ having the same coefficients as h. Write $x_s = \xi_s + i\eta_s$, where ξ and η are real. Then H is the sum of $\sum b_{st} \xi_s \xi_t$, which by hypothesis is positive unless the ξ's are all zero, and $\sum b_{st} \eta_s \eta_t$, which is positive unless the η's are all zero. Hence H is positive unless the x's are all zero. Thus λ_1 is real by the first case.

Thus $f - \lambda_1 h$ is a real quadratic form or a Hermitian form in the respective cases. Since its determinant is zero, it can be reduced by Theorem 7 to a real quadratic or Hermitian form $F(y_2, \ldots, y_n)$ lacking one variable y_1 by a non-singular linear transformation which is real in the quadratic case. Let the same transformation replace h by H. Hence it replaces

$$f - \lambda h \equiv f - \lambda_1 h + (\lambda_1 - \lambda)h$$

by

(22) $\quad f - \lambda h \equiv F(y_2, \ldots, y_n) + (\lambda_1 - \lambda)H(y_1, \ldots, y_n).$

In the positive form H of rank n, the coefficient of $y_1 \tilde{y}_1$ is not zero, since otherwise H would vanish when $y_1 = 1$, $y_k = 0$ $(k > 1)$.

Hence by the last part of the proof of Theorem 7, there exists a non-singular linear transformation which leaves each y_k $(k > 1)$ unaltered and replaces H by a form $c_1 g_1 \tilde{g}_1 + \psi(y_2, \ldots, y_n)$, where c_1 is positive. This transformation, which is real in the quadratic case, replaces the second member of (22) by

$$F(y_2, \ldots, y_n) + (\lambda_1 - \lambda)c_1 g_1 \tilde{g}_1 + (\lambda_1 - \lambda)\psi(y_2, \ldots, y_n).$$

Write ϕ for $F + \lambda_1 \psi$. Hence

$$f - \lambda h \equiv \phi(y_2, \ldots, y_n) + (\lambda_1 - \lambda)c_1 g_1 \tilde{g}_1 - \lambda\psi(y_2, \ldots, y_n).$$

Equating the terms free of λ and those containing λ, we have

$$f \equiv \lambda_1 c_1 g_1 \tilde{g}_1 + \phi(y_2, \ldots, y_n), \qquad h \equiv c_1 g_1 \tilde{g}_1 + \psi(y_2, \ldots, y_n).$$

By the hypothesis on h, evidently ψ is positive and of rank $n-1$. Hence the discussion made for f, h, and λ_1 is applicable to ϕ, ψ, and λ_2 and leads to the segregation of terms $\lambda_2 c_2 g_2 \tilde{g}_2$ and $c_2 g_2 \tilde{g}_2$. Repetitions of this argument show that f and h are reducible, by a non-singular transformation which is real in the quadratic case, to forms

$$f \equiv \sum \lambda_i c_i g_i \tilde{g}_i, \qquad h \equiv \sum c_i g_i \tilde{g}_i,$$

in which the c_i are all positive. The transformation

$$\xi_i = c_i^{\frac{1}{2}} g_i \qquad (i = 1, \ldots, n)$$

is real and non-singular, and yields the canonical forms given in Theorems 10, 11.

As a special case of Theorem 10, we may take h to be $x_1^2 + \cdots + x_n^2$. Then the linear transformation leaves h unaltered and is called *orthogonal*. Similarly, if in Theorem 11 we take h to be $x_1 \bar{x}_1 + \ldots + x_n \bar{x}_n$, the transformation leaves h unaltered and is called *unitary*. Theorems 10 and 11 therefore imply (and are really no more general than) the following two:

THEOREM 12. *If f is a real quadratic form of matrix A, the roots $\lambda_1, \ldots, \lambda_n$ of its characteristic equation $|A - \lambda I| = 0$ are all real, and f can be reduced to $\sum \lambda_i \xi_i^2$ by a real orthogonal transformation.*

THEOREM 13. *If f is any Hermitian form of matrix A, the roots $\lambda_1, \ldots, \lambda_n$ of its characteristic equation $|A - \lambda I| = 0$ are all real, and f can be reduced to $\sum \lambda_i \xi_i \bar{\xi}_i$ by a unitary transformation.*

EXERCISES

1. A linear transformation with matrix B is orthogonal if and only if $B'B = I$ and is unitary if and only if $B'\bar{B} = I$. Hint: Take $A = A_1 = I$ in §§33, 34. .

2. The determinant of an orthogonal transformation is ± 1; that of a unitary transformation has the absolute value 1.

§38] RANK OF SYMMETRIC OR HERMITIAN MATRIX

3. The product of two orthogonal transformations and the inverse of one are orthogonal. Likewise for unitary.

4. If a linear transformation has two of the properties (i) is real, (ii) is orthogonal, (iii) is unitary, then it has the third property.

5. Every real orthogonal transformation on x and y is of the form

$$x' = x \cos \alpha - y \sin \alpha, \qquad y' = \pm (x \sin \alpha + y \cos \alpha).$$

6. The matrix of every unitary transformation of determinant unity on two variables is of the form

$$\begin{pmatrix} a & -\bar{c} \\ c & \bar{a} \end{pmatrix}.$$

7. The transformation from one system of rectangular coordinates to another having the same origin is orthogonal. Then by means of Theorem 12 show that the equation of any real quadric surface can be reduced to a form in which the terms of the second degree are $ax^2 + by^2 \pm cz^2$, $ax^2 \pm by^2$, or ax^2, where a, b, c are positive.

8. In the canonical form (20), there are p positive and $N = r - p$ negative coefficients. The number $p - N$ is called the *signature*. Two real quadratic forms are equivalent under real transformation if and only if they have the same rank and same signature.

38. Rank of a symmetric or Hermitian matrix.

THEOREM 14. *If a certain r-rowed determinant d of a matrix is not zero and if all $(r + 1)$-rowed determinants having d as a first minor are zero, then the matrix is of rank r.*

Note that this theorem shortens the work of finding the rank r, since we need not prove that all $(r + 1)$-rowed determinants are zero, but merely those having d as a first minor.

After interchanges of rows or columns or both, we obtain a matrix

$$A = \begin{bmatrix} a_{11} & \cdots & a_{1n} \\ \cdots & \cdots & \cdots \\ a_{m1} & \cdots & a_{mn} \end{bmatrix}$$

in which

$$\Delta = \begin{vmatrix} a_{11} & \cdots & a_{1r} \\ \cdots & \cdots & \cdots \\ a_{r1} & \cdots & a_{rr} \end{vmatrix} \neq 0.$$

The theorem will evidently follow if proved for matrix A and determinant Δ. Write

$$l_i = a_{i1} y_1 + \cdots + a_{in} y_n,$$
$$\lambda_i = a_{i1} y_1 + \cdots + a_{ir} y_r, \qquad (i = 1, \ldots, m).$$

For $s > r$, $t > r$, the determinant of order $r + 1$ of the forms

$$\lambda_i + a_{it} y_t \qquad (i = 1, \ldots, r, s)$$

is zero by hypothesis. Hence these $r + 1$ linear forms are linearly dependent by Theorem 9 of Ch. III, so that

$$(23) \qquad \lambda_s + a_{st} y_t = \sum_{i=1}^{r} c_i (\lambda_i + a_{it} y_t).$$

Since $\Delta \neq 0$, we may solve equations $\lambda_i = a_{i1} y_1 + \cdots$ and express $y_1, \ldots, y_r$ linearly in terms of $\lambda_1, \ldots, \lambda_r$. Substitution in λ_s gives

$$\lambda_s = b_1 \lambda_1 + \cdots + b_r \lambda_r.$$

Inserting this into (23) and noting that $\lambda_1, \ldots, \lambda_r, y_t$ are linearly independent since $y_1, \ldots, y_r, y_t$ are, we see that

$$c_i = b_i \qquad (i = 1, \ldots, r), \qquad a_{st} = \sum_{i=1}^{r} b_i a_{it}.$$

Then

$$l_s \equiv \lambda_s + \sum_{t=r+1}^{n} a_{st} y_t = \sum_{i=1}^{r} b_i \left(\lambda_i + \sum_{t=r+1}^{n} a_{it} y_t \right) \equiv \sum_{i=1}^{r} b_i l_i.$$

[§ 38] RANK OF SYMMETRIC OR HERMITIAN MATRIX

Since $l_{r+1}, \ldots, l_m$ are therefore linear functions of $l_1, \ldots, l_r$, which are linearly independent, matrix A of the l's is of rank r by Theorem 13 of Ch. III.

We shall next prove theorems which enable us to find the rank of a matrix A having the property $A' = \tilde{A}$ by an examination of its principal minors only. A *principal minor* of any square matrix A is the determinant of the matrix obtained from A by deleting the same rows as columns; in other words, its diagonal elements all lie in the diagonal of A.

THEOREM 15. *The rank of a symmetric or Hermitian matrix A is r if an r-rowed principal minor p_r is not zero and if zero is the value of every principal minor obtained by annexing to p_r one row and the same column of A, and also of every principal minor obtained by annexing two rows and the same two columns.*

Without loss of generality we may assume that the elements of p_r lie in the first r rows and first r columns of $A = (a_{ij})$. Let g_{ij} denote the determinant obtained from p_r by annexing the ith row and jth column of A. Our theorem will follow from Theorem 14 if we prove that $g_{ij} = 0$ for all distinct values of i and j. Let c denote the determinant obtained from p_r by annexing the ith and jth rows and the ith and jth columns of A. Let A_{ij} denote the cofactor of a_{ij} in c. By a standard theorem on determinants,

$$(24) \qquad c\, p_r = \begin{vmatrix} A_{ii} & A_{ij} \\ A_{ji} & A_{jj} \end{vmatrix} \equiv g_{ij} g_{ii} - g_{ij} g_{ji}.$$

By hypothesis, g_{ii}, g_{jj}, and c are zero. Since $A' = \tilde{A}$, if we interchange the rows and columns of g_{ji}, we get $\tilde{g}_{ij}$. Hence

$$0 = g_{ij} g_{ji} = g_{ij} \tilde{g}_{ij}, \qquad g_{ij} = 0.$$

THEOREM 16. *If all $(r + 1)$-rowed principal minors and all $(r + 2)$-rowed principal minors of a symmetric or Hermitian matrix A are zero, the rank of A is r or less.*

This is true when $r = 0$, since $a_{ii} = a_{jj} = 0$ and $a_{ii} a_{jj} - a_{ij} a_{ji} = 0$ imply that $a_{ij} \tilde{a}_{ij} = 0$ and hence that every element a_{ij} is zero, whence A is of rank zero.

Proceeding by induction on r, we assume that the theorem is true when $r = k$, that is, if all $(k + 1)$-rowed principal minors and all $(k + 2)$-rowed principal minors are zero, the rank of A is $\leq k$. To prove the theorem when $r = k + 1$, let all $(k + 2)$-rowed principal minors and all $(k + 3)$-rowed principal minors be zero. Then if all $(k + 1)$-rowed principal minors are zero, the rank is $\leq k$ by the assumption just made. But if some $(k + 1)$-rowed principal minor is not zero, the rank is $k + 1$ by Theorem 15. In either case, the rank is $\leq k + 1$, and the induction is complete.

THEOREM 17. *Any symmetric or Hermitian matrix of rank r ($r > 0$) contains an r-rowed principal minor which is not zero.*[1]

By the definition of rank r, all $(r + 1)$-rowed principal minors are zero. If all r-rowed principal minors were zero, the rank would be $\leq r - 1$ by Theorem 16.

EXERCISES

1. By means of Theorem 16 prove that 2 is the rank of

$$\begin{pmatrix} -6 & -5 & 1 & -2 \\ -5 & 56 & 4 & 11 \\ 1 & 4 & 0 & 1 \\ -2 & 11 & 1 & 2 \end{pmatrix}.$$

2. Call a matrix A *skew-symmetric* if $A' = -A$. Its determinant is zero if of odd order. Extend Theorems 15–17 to skew-symmetric matrices. Using Theorem 17, prove that the rank of a skew-symmetric matrix is always even.

3. If A is a real skew-symmetric matrix, $P = iA$ is Hermitian. Hence give a more direct proof of Ex. 2. The characteristic equation $|A - \lambda I| = 0$ of any real skew-symmetric matrix has only purely imaginary roots.

---

[1] For elementary, direct proofs, see Dickson, Wedderburn, and Bliss, Annals of Math., (2), 15, 1913, 27, 29; 16, 1914, 43.

§ 38] RANK OF SYMMETRIC OR HERMITIAN MATRIX

4. By an *alternate* bilinear form is meant

$$\alpha = \sum_{k,j=1}^{n} a_{kj} x_k y_j, \qquad a_{jk} = -a_{kj}.$$

Show that, if $a_{12} \neq 0$,

$$x_2 = a_{12}^{-1}(X_2 - \sum_{j=3}^{n} a_{1j} X_j), \qquad x_i = X_i \qquad (i = 1, 3, 4, \ldots, n)$$

and the cogredient transformation on the y's replace α by $X_1 Y_2 - Y_1 X_2 + \phi$, where ϕ is an alternate bilinear form in X_i, $Y_i (i = 2, \ldots, n)$. Treating ϕ as we did α, prove that any alternate bilinear form α with coefficients in any field F is equivalent under cogredient non-singular transformations with coefficients in F to

$$\xi_1 \eta_2 - \xi_2 \eta_1 + \xi_3 \eta_4 - \xi_4 \eta_3 + \cdots + \xi_{2s-1} \eta_{2s} - \xi_{2s} \eta_{2s-1}.$$

5. By an alternate Hermitian bilinear form is meant a form α in Ex. 4 in which now $a_{jk} = -\bar{a}_{kj}$. If its matrix is A, show that $B = iA$ is a Hermitian bilinear form. What is therefore the canonical form of α?

THEOREM 18. *If M is a symmetric or Hermitian matrix of rank r ($r > 0$), we may derive a **regular** symmetric or Hermitian matrix $A = (a_{ij})$ by a permutation of the rows of M and the same permutation of the corresponding columns such that no consecutive two of the numbers*

$$(25) \quad p_0 = 1, \; p_1 = a_{11}, \; p_2 = \begin{vmatrix} a_{11} & a_{12} \\ a_{21} & a_{22} \end{vmatrix}, \ldots, p_r = \begin{vmatrix} a_{11} & \cdots & a_{1r} \\ \cdots & & \cdots \\ a_{r1} & \cdots & a_{rr} \end{vmatrix}$$

are zero, where the p_i are principal minors and $p_r \neq 0$.

Postponing the case in which all the elements of the diagonal of M are zero, let the ith element be not zero. After interchanging the first and ith rows and first and ith columns of M, we obtain a symmetric or Hermitian matrix M_1 having $a_{11} \neq 0$.

Postponing the case in which zero is the value of every two-rowed principal minor, containing a_{11}, of M_1, let j be such that the two-rowed minor obtained by deleting all rows and columns except those numbered 1 and j is not zero. After interchanging the second and jth rows and second and jth columns of M_1, we obtain a symmetric or Hermitian matrix M_2 having $a_{11} \neq 0$, $p_2 = a_{11} a_{22} - a_{12} a_{21} \neq 0$.

Consider those three-rowed principal minors of M_2 which have p_2 as a first minor and proceed as before.

In this manner we will ultimately obtain a symmetric or Hermitian matrix M_r in which no one of the numbers (25) is zero unless we meet one of the postponed cases. The general such case is as follows. Let each of the numbers $p_1 = a_{11}, p_2, \ldots, p_k$ be not zero, where $k < r$, while zero is the value of every $(k+1)$-rowed principal minor having p_k as a first minor. By Theorem 15, not every $(k+2)$-rowed principal minor having p_k as a second minor is zero, since otherwise the rank of M would be $k < r$. After a permutation of the rows numbered $k+1, k+2, \ldots, n$ and the same permutation of the corresponding columns, we have $p_{k+2} \neq 0$, $p_{k+1} = 0$.

To prove the theorem by induction, assume that we have reached a matrix B in which no consecutive two of the numbers $p_1, \ldots, p_s$ are zero, while $p_s \neq 0$, $s < r$. If not all $(s+1)$-rowed principal minors (of B) having p_s as a first minor are zero, we permute rows and also columns numbered $s+1, \ldots, n$ and obtain a matrix with also $p_{s+1} \neq 0$. In the contrary case, we proceed as in the preceding paragraph with $k = s$ and obtain a matrix with also $p_{s+2} \neq 0$, $p_{s+1} = 0$.

39. Kronecker's[1] method of reduction.

THEOREM 19. *Consider a quadratic or Hermitian form*

(26) $$Q = \tilde{Q} = \sum_{i, j=1}^{n} a_{ij} x_i \tilde{x}_j, \qquad a_{ji} = \tilde{a}_{ij},$$

of rank r $(0 < r < n)$ with coefficients in the field F. After a permutation of the x's we may assume from Theorem 17 that $|a_{ij}| \neq 0$

---

[1] Werke, I, 166, 357. Cf. Dickson, Trans. Amer. Math. Soc., 10, 1909, 129, who gave applications to modular invariants.

$(i, j = 1, \ldots, r)$. Then we may express the x's in terms of new variables y's by a linear transformation with coefficients in F of determinant unity such that $x_i = y_i$ $(i = r + 1, \ldots, n)$ and such that Q becomes

$$Q_1 = \sum_{i, j=1}^{r} a_{ij} y_i \tilde{y}_j,$$

whose terms have the same coefficients as the corresponding terms of Q.

According as $\tilde{x}$ denotes x or $\bar{x}$, write $t = 2$ or 1; then

$$\frac{\partial Q}{\partial x_i} = t \sum_{j=1}^{n} a_{ij} \tilde{x}_j \qquad (i = 1, \ldots, n).$$

Since the determinant of these n linear forms is of rank r, Theorems 10 and 13 of Ch. III show that (the first) r of them are linearly independent, while the remaining $n - r$ are linearly dependent on those r:

(27) $\qquad \dfrac{\partial Q}{\partial x_k} = \sum_{i=1}^{r} c_{ki} \dfrac{\partial Q}{\partial x_i} \qquad (k = r + 1, \ldots, n),$

where the c_{ki} are numbers of F. Consider the transformation

(28) $\quad \begin{aligned} x_i &= y_i - \sum_{k=r+1}^{n} c_{ki} y_k \quad (i = 1, \ldots, r), \\ x_k &= y_k \quad (k = r + 1, \ldots, n) \end{aligned}$

of determinant unity and the implied relations

(29) $\quad \tilde{x}_i = \tilde{y}_i - \sum_{k=r+1}^{n} \tilde{c}_{ki} \tilde{y}_k \quad (i = 1, \ldots, r),$

$$\tilde{x}_k = \tilde{y}_k \quad (k = r + 1, \ldots, n).$$

Hence

$$\frac{\partial Q}{\partial y_k} = \sum_{i=1}^{n} \frac{\partial Q}{\partial x_i} \frac{\partial x_i}{\partial y_k} = \frac{\partial Q}{\partial x_k} - \sum_{i=1}^{r} c_{ki} \frac{\partial Q}{\partial x_i} = 0$$

$$(k = r + 1, \ldots, n),$$

and, since $\tilde{Q} = Q$,

$$\frac{\partial Q}{\partial \tilde{y}_k} = \frac{\partial \tilde{Q}}{\partial \tilde{y}_k} = \sum_{i=1}^{n} \frac{\partial \tilde{Q}}{\partial \tilde{x}_i} \frac{\partial \tilde{x}_i}{\partial \tilde{y}_k} = \frac{\partial \tilde{Q}}{\partial \tilde{x}_k} - \sum_{i=1}^{r} \tilde{c}_{ki} \frac{\partial \tilde{Q}}{\partial \tilde{x}_i} = 0$$

$$(k = r+1, \ldots, n),$$

by (27) and the relation obtained from it by placing the symbol $\sim$ over every number of it. Thus Q is transformed by (28) and (29) into a form Q_1 lacking y_k and $\tilde{y}_k$ ($k = r+1, \ldots, n$). Hence Q_1 is derived from Q by writing $x_i = y_i$, $\tilde{x}_i = \tilde{y}_i$ ($i = 1, \ldots, r$), $x_k = \tilde{x}_k = 0$ ($k = r+1, \ldots, n$), so that Q_1 is the form given in the theorem.

In the following two theorems, A_{ij} denotes the cofactor of a_{ij} in the determinant a of the form (26).

Theorem 20. *If $A_{nn} \neq 0$, we may express the x's in terms of new variables $y_1, \ldots, y_n$ by a linear transformation with coefficients in F of determinant unity such that $x_n = y_n$ and such that (26) becomes*

$$(30) \quad \sum_{i,\,j=1}^{n-1} a_{ij} y_i \tilde{y}_j + \frac{a}{A_{nn}} y_n \tilde{y}_n.$$

In proof, consider the form

$$\phi = \sum_{i,\,j=1}^{n} a_{ij} x_i \tilde{x}_j - \frac{a}{A_{nn}} x_n \tilde{x}_n.$$

Its determinant is derived from a by replacing a_{nn} by $a_{nn} - a/A_{nn}$ and hence is equal to the sum of a and a determinant obtained from a by replacing the last row by $0, \ldots, 0, -a/A_{nn}$. Hence the determinant of ϕ is

$$a - A_{nn} \cdot a/A_{nn} = 0.$$

Thus ϕ is of rank $n - 1$. Theorem 19 therefore shows the existence of a transformation of the type specified in Theorem 20 which replaces ϕ by

§ 39] KRONECKER'S REDUCTION

$$\sum_{i,\,j=1}^{n-1} a_{ij} y_i \tilde{y}_j$$

and $x_n \tilde{x}_n$ by $y_n \tilde{y}_n$, and hence replaces (26) by (30).

THEOREM 21. *If $A_{nn} = 0$, $A_{n-1\,n-1} = 0$, $A_{n\,n-1} \neq 0$, we may express the x's in terms of new variables $y_1, \ldots, y_n$ by a linear transformation with coefficients in F of determinant unity such that $x_{n-1} = y_{n-1}$, $x_n = y_n$, and such that (26) becomes*

$$(31) \quad \sum_{i,\,j=1}^{n-2} a_{ij} y_i \tilde{y}_j + g y_{n-1} \tilde{y}_n + \tilde{g} y_n \tilde{y}_{n-1}, \quad g = a/\tilde{A}_{nn-1}.$$

Let b denote the determinant obtained by deleting the last two rows and last two columns of a. Then, as in (24),

$$(32) \quad ab = \begin{vmatrix} A_{n-1\,n-1} & A_{n-1\,n} \\ A_{nn-1} & A_{nn} \end{vmatrix} = -A_{nn-1} \tilde{A}_{nn-1} \neq 0.$$

Consider the form

$$\psi = \sum_{i,\,j=1}^{n} a_{ij} x_i \tilde{x}_j - g x_{n-1} \tilde{x}_n - \tilde{g} x_n \tilde{x}_{n-1}.$$

Its matrix is

$$M = \begin{pmatrix} a_{11} & \cdots & a_{1\,n-1} & & a_{1n} \\ \cdots\cdots\cdots\cdots\cdots\cdots\cdots\cdots\cdots\cdots\cdots \\ a_{n-1\,1} & \cdots & a_{n-1\,n-1} & & a_{n-1\,n} - g \\ a_{n1} & \cdots & a_{n\,n-1} - \tilde{g} & & a_{nn} \end{pmatrix}.$$

The determinant obtained by deleting the last two rows and last two columns is b, which is not zero by (32) and is the p_r of Theorem 15. Next, the principal minor of M obtained by deleting the last row and last column is $A_{nn} = 0$, while that obtained by deleting the $(n-1)$th row and $(n-1)$th column is $A_{n-1\,n-1} = 0$ [they are the only $(n-1)$-rowed principal minors having b as a first minor]. Finally, if d is the determinant of M, the theorem which gave (32) implies

$$db = \begin{vmatrix} A_{n-1\;n-1} & A_{n-1\;n} + \tilde{g}b \\ A_{n\;n-1} + gb & A_{nn} \end{vmatrix} = -t\tilde{t},$$

where $t = A_{n\;n-1} + gb$, since $\tilde{b} = b$. Replacing g by its value in (31) and using (32), we get

$$\tilde{A}_{n\;n-1}\,t = b(g\tilde{A}_{n\;n-1} - a) = 0, \qquad t = 0.$$

Hence $db = 0$. Thus $d = 0$. Hence M is of rank $n - 2$ by Theorem 15. Hence there exists by Theorem 19 a transformation of the type specified in Theorem 21 which replaces ψ by

$$\sum_{i,\,j=1}^{n-2} a_{ij}\,y_i\,\tilde{y}_j,$$

and therefore replaces (26) by (31).

If we apply to (31) the transformation

$$y_i = x_i\,(i = 1, \ldots, n-2), \quad y_{n-1} = x_{n-1} - \tfrac{1}{2}\tilde{g}x_n,$$
$$y_n = (1/\tilde{g})x_{n-1} + \tfrac{1}{2}x_n,$$

of determinant unity, we obtain

$$(33) \quad \sum_{i,\,j=1}^{n-2} a_{ij}\,x_i\,\tilde{x}_j + 2x_{n-1}\,\tilde{x}_{n-1} - \tfrac{1}{2}g\tilde{g}x_n\,\tilde{x}_n, \qquad g = a/\tilde{A}_{nn-1}.$$

Let Q be a form (26) of rank r with coefficients in a field F. We can find by means of Theorems 19–21 a linear transformation of determinant unity with coefficients in F which replaces Q by

$$(34) \qquad g_1\,x_1\,\tilde{x}_1 + \cdots + g_r\,x_r\,\tilde{x}_r \qquad (g_i = \tilde{g}_i \neq 0),$$

in which the g's are obtained explicitly in terms of the coefficients a_{ij} of Q by formulas which differ in the various subcases necessary to consider. We first pass at once to the equivalent form Q_1 of Theorem 19.

(i) If not all $(r-1)$-rowed principal minors of the determinant a of Q_1 are zero, we may permute the x's and assume that the cofactor A_{rr} of a_{rr} in a is not zero. We obtain the equivalent form (30) of Theorem 20 with $n = r$, having the segregated term $(a/A_{rr})x_r \tilde{x}_r$.

(ii) But if all $(r-1)$-rowed principal minors of a are zero, the cofactor A_{ij} of at least one element $a_{ij}(i \neq j)$ is not zero, since the rank is r. Permuting the x's, we may assume that $A_{r\,r-1} \neq 0$. We obtain the equivalent form (33) with $n = r$ having the last two terms segregated.

Treating the unreduced part (under the summation sign) as in (i) or (ii), we finally obtain (34) in which the coefficients g_i are expressed in terms of determinants of the matrix of the given form Q.

For example, if the first of the two alternatives (i) and (ii) holds true at each stage of the reduction, we obtain the equivalent form (34) with the coefficients

$$(35) \qquad \frac{p_1}{p_0}, \frac{p_2}{p_1}, \frac{p_3}{p_2}, \ldots, \frac{p_{r-1}}{p_{r-2}}, \frac{p_r}{p_{r-1}},$$

where $p_0 = 1$ and the remaining p's are the special principal minors (25) and each is not zero in the present case.

40. Number of positive coefficients in the canonical form. By Theorem 18, the variables of Q may be so permuted that no consecutive two of the numbers $p_0, p_1, \ldots, p_r$ are zero, while $p_r \neq 0$; call such a form *regular*.

Let the field F be real when Q is a quadratic form. When Q is an Hermitian form, let F be a field obtained from a real field S by the adjunction of $\nu^{\frac{1}{2}}$, where ν is in S and is negative. Then if $x = a + b\nu^{\frac{1}{2}}$, where a and b are in S, $x\tilde{x} = a^2 - \nu b^2$ is positive unless $x = 0$. In either case, $p_i = \tilde{p}_i$ is real, while the coefficients of the last two segregated terms of (33) are of opposite signs.

Let P be the number of permanences of sign and V the number of variations of sign in the sequence $p_0, p_1, \ldots, p_r$, with the agreement that a vanishing p_i is counted as positive or negative at pleasure, but must be counted.

Let p be the number of positive coefficients and n the number of negative coefficients in the equivalent form (34).

If each $p_i \neq 0$, these coefficients are given by (35). For each permanence (variation) of sign, two consecutive p's are of like (opposite) signs and their quotient is a positive (negative) coefficient (35), and conversely. Hence $p = P$, $n = V$.

This result holds true also when any p_i is zero. For, p_{i-1} and p_{i+1} are then each $\neq 0$ (since Q is regular) and have opposite signs by (32) with $n = i + 1$, $a = p_{i+1}$, $b = p_{i-1}$, $A_{nn} = p_i$. By the final two terms of (33), we now have two coefficients of opposite signs instead of the two fractions (35) which involve p_i. We have $P = V = 1$ for the sequence p_{i-1}, $p_i = 0$, p_{i+1} whose signs are $+ \pm -$ or $- \pm +$ by the agreement concerning zero. In other words, by ignoring the permanences and variations of sign in this sequence of three terms, we reduce P, V, p, and n each by 1. After all such deletions due to vanishing p_i's, we saw that $p = P$, $n = V$. Hence the latter hold also before the deletions.

THEOREM 22. *Within any real field, any quadratic form of rank r is equivalent to one of type $g_1 x_1^2 + \cdots + g_r x_r^2$. Within a field obtained from any real field S by the adjunction of the square root of a negative number of S, any Hermitian form of rank r is equivalent to one of type $g_1 x_1 \bar{x}_1 + \cdots + g_r x_r \bar{x}_r$, with each g_i in S. Then the number of positive g's is the number of permanences of sign in the sequence $p_0, p_1, \ldots, p_r$, no consecutive two of which are zero, as may be assumed after a suitable permutation of the x's.*

For example, let $Q = x_1^2 + 2x_2 x_3 + x_4^2$. Then $p_0 = 1$, $p_1 = 1$, $p_2 = 0$, $p_3 = -1$, $p_4 = -1$, $P = 3$, $V = 1$. Since Q is equivalent to $x_1^2 + 2(\xi^2 - \eta^2) + x_4^2$, we have $p = 3$, $n = 1$.

CHAPTER V

THEORY OF LINEAR TRANSFORMATIONS,
INVARIANT FACTORS AND ELEMENTARY DIVISORS

Weierstrass's elementary divisors and the related invariant factors are here introduced in a natural manner in connection with the classic and a new rational canonical form of linear transformations, respectively.

41. Rational canonical form of a linear transformation.[1] Let $w_1, \ldots, w_n$ be independent variables. Let S be any linear transformation

$$S: \quad W_i = \sum_{j=1}^{n} a_{ij} w_j \qquad (i = 1, \ldots, n)$$

with coefficients in any field F. We do not exclude the case in which the determinant of S is zero, so that the matrix of the coefficients of S is perfectly general.

Let x_1 be any homogeneous linear function of $w_1, \ldots, w_n$ whose coefficients belong to F and are not all zero. Let X_1 denote the corresponding function of $W_1, \ldots, W_n$. If $X_1 = cx_1$, where c is in F, we have (1) for $a = 1$, with the understanding that the final equation alone is retained. But if $X_1 \neq cx_1$, we write x_2 for X_1. Let X_2 denote the same function of the W_i that x_2 is of the w_i. Then if X_2 is a linear function of x_1 and x_2, we have (1) for $a = 2$; in the contrary case we write x_3 for X_2 and have three linearly independent functions x_1, x_2, x_3. Proceeding in this manner, we obtain

---

[1] For the case in which the determinant is not zero, S. Lattès, Annales Fac. Sc. Toulouse, (3), 6, 1914, 1–84. Also W. Krull, Über Begleitmatrizen und Elementarteilertheorie, Freiburg thesis, 1921. In Leipzig. Berichte, 69, 1917, 325–35, G. Kowalewski employed points, instead of linear functions, and obtained a canonical form whose matrix is the transpose of that in the text. The text presents the theory as found independently by the author several years ago.

a *chain* of linearly independent functions $x_1, x_2, \ldots, x_a$ with coefficients in F such that S implies

(1) $\quad X_1 = x_2, \ X_2 = x_3, \ \ldots, \ X_{a-1} = x_a, \ X_a = [x_1, \ldots, x_a],$

where the bracket denotes a homogeneous linear function of $x_1, \ldots, x_a$ with coefficients in F. We call a the *length* of the chain, and x_1 its *leader*.

THEOREM 1. *By the introduction of new variables which are linearly independent homogeneous linear functions of the initial variables w_i with coefficients in the field F, any linear transformation S with coefficients in F may be reduced to a canonical form defined by* (1) *and*

(2) $\quad Y_1 = y_2, \ Y_2 = y_3, \ \ldots, \ Y_{b-1} = y_b, \ Y_b = [y_1, \ldots, y_b];$

(3) $\quad Z_1 = z_2, \ Z_2 = z_3, \ \ldots, \ Z_{c-1} = z_c, \ Z_c = [z_1, \ldots, z_c];$

etc., where a is the maximal length of all possible chains, b is the maximal length of a chain whose leader is linearly independent of $x_1, \ldots, x_a$, and c is the maximal length of a chain whose leader is linearly independent of $x_1, \ldots, x_a, y_1, \ldots, y_b$.

If $a = 1$, S is $W_i = a_{ii} w_i$ $(i = 1, \ldots, n)$, and is in canonical form with n chains. Henceforth let $a > 1$ and select a function x_1 which is the leader of a chain of maximal length a, so that we have (1). If $n = a$, the theorem is proved.

For $n > a$, let u_1 be any homogeneous linear function of the initial variables w_i with coefficients in F which is linearly independent of $x_1, \ldots, x_a$. Let U_1 denote the corresponding function of the W_i. If $U_1 = [x_1, \ldots, x_a, u_1]$, we have (4) for $b = 1$. In the contrary case, write u_2 for U_1. Then if $U_2 = [x_1, \ldots, x_a, u_1, u_2]$, we have (4) for $b = 2$. In the contrary case, write u_3 for U_2. Proceeding in this manner, we get

(4) $\quad U_1 = u_2, \ U_2 = u_3, \ \ldots, \ U_{b-1} = u_b, \ U_b = l + [u_1, \ldots, u_b],$

§ 41]   RATIONAL CANONICAL FORM   91

where $l = [x_1, \ldots, x_a]$, and $x_1, \ldots, x_a, u_1, \ldots, u_b$ are linearly independent. If $l \equiv 0$, (4) is the desired result (2) with $y_i = u_i$.

Next, let $l \not\equiv 0$. Then $l + [u_1, \ldots, u_b]$ is linearly independent of $u_1, \ldots, u_b$, so that u_1 is the leader of a chain whose length exceeds b. Hence $b < a$. Introduce the new variables

$$y_1 = u_1 + f, \qquad y_2 = u_2 + f_1, \ldots, y_b = u_b + f_{b-1},$$

where f is a homogeneous linear function of $x_1, \ldots, x_a$ with coefficients in F, and S replaces f by f_1, f_1 by f_2, etc. We can choose [1] f so that (2) follows.

To obtain the third chain, let v_1 be any homogeneous linear function of the initial variables w_i with coefficients in F such that $x_1, \ldots, x_a, y_1, \ldots, y_b, v_1$ are linearly independent. Let V_1 denote the corresponding function of the W_i. If $V_1 = [x_1, \ldots, x_a, y_1, \ldots, y_b, v_1]$, we have (5) for $c = 1$. In the contrary case, write v_2 for V_1. Then if $V_2 = [x_1, \ldots, x_a, y_1, \ldots, y_b, v_1, v_2]$, we have (5) for $c = 2$; in the contrary case, write v_3 for V_2. Proceeding in this manner, we get

(5) $V_1 = v_2, \quad V_2 = v_3, \ldots, V_{c-1} = v_c, \quad V_c = l + m,$

where

$$l = [x_1, \ldots, x_a, y_1, \ldots, y_b], \qquad m = s_1 v_1 + \cdots + s_c v_c.$$

Here $x_1, \ldots, x_a, y_1, \ldots, y_b, v_1, \ldots, v_c$ are linearly independent, and the s_i are numbers of F. If $l \equiv 0$, we have the desired result (3) with $z_i = v_i$.

Next, let $l \not\equiv 0$. Then $l + m$ is linearly independent of $v_1, \ldots, v_c$, so that v_1 is the leader of a chain whose length exceeds c. Since v_1 is linearly independent of $x_1, \ldots, x_a$ and since b is the maximum length of a chain whose leader is linearly independent of $x_1, \ldots, x_a$, we have $c < b$. Introduce the new variables

---

[1] The proof is like that concerning (6) with the y_i suppressed and z_i replaced by y_i, c by b, and, beginning with (8), also b by a. We prefer to omit the details here and treat the more typical derivation of the third chain.

(6) $\quad z_1 = v_1 + f, \quad z_2 = v_2 + f_1, \ldots, z_c = v_c + f_{c-1}$,

where $f = f_0$ is a homogeneous linear function of $x_1, \ldots, x_a$, $y_1, \ldots, y_b$ with coefficients in F, while S replaces f by f_1, f_1 by f_2, etc. Let Z_j denote the same function of the W_i that z_j is of the w_i. Hence

(7) $\quad Z_1 = z_2, Z_2 = z_3, \ldots, Z_{c-1} = z_c, Z_c = l_1 + \sum_{i=1}^{c} s_i z_i$,

where

$$l_1 = l + f_c - \sum_{i=1}^{c} s_i f_{i-1}.$$

We can choose f so that l_1 becomes a homogeneous linear function of $x_i, y_i (i = 1, \ldots, c)$. The proof is simpler by successive steps. First, if k is the coefficient of x_a in l, take $f = -kx_{a-c}$, whence $f_c = -kx_a$, and l_1 lacks x_a. Since (7) is of the form (5), we may therefore assume that l itself lacks x_a. Then if t is the coefficient of x_{a-1} in l, take $f = -tx_{a-c-1}$, whence $f_c = -tx_{a-1}$, and l_1 lacks both x_{a-1} and x_a. The last of these steps employs $f = -gx_1$ to delete x_{c+1} from l_1. Similarly, we can choose the coefficients of $y_{b-c}, y_{b-c-1}, \ldots, y_1$ of f to delete $y_b, y_{b-1}, \ldots, y_{c+1}$ from l_1.

If the resulting function l_1 is identically zero, (7) is the desired result (3). In the contrary case, write

$$l_i = t_1 x_i + \cdots + t_c x_{c+i-1} + m_1 y_i + \cdots + m_c y_{c+i-1}$$
$$(i = 1, \ldots, b - c + 1).$$

Since we proved that l_1 is given by this formula with $i = 1$, the t's and m's are not all zero. Let L_i denote the corresponding function of the X's and Y's. Hence

(8) $\quad L_i = l_{i+1} \qquad (i = 1, \ldots, b - c)$.

Evidently $l_i (i = 1, \ldots, b - c + 1)$ are linearly independent linear functions of $x_j, y_j (j = 1, \ldots, b)$. By (7) and (8), the chain with the leader z_1 has the successive terms

$$z_1, \ldots, z_c, \quad z_{c+j} = l_j + \sum_{i=1}^{c} s_i z_{i+j-1} \quad (j = 1, \ldots, b - c + 1),$$

and possibly further terms. By (6), the $b + 1$ terms z are linearly independent. Since the leader z_1 is linearly independent of $x_1, \ldots, x_a$, this result contradicts the definition of b.

42. Theorem 2. *In a canonical transformation the characteristic determinant of the partial transformation on any chain is divisible by that of the next chain.*

For definiteness let the chains be the second and the third, and let the partial transformations on them be

(9) $\quad Y_1 = y_2, \; Y_2 = y_3, \ldots, Y_{b-1} = y_b, \; Y_b = \sum_{i=1}^{b} b_i y_i;$

(10) $\quad Z_1 = z_2, \; Z_2 = z_3, \ldots, Z_{c-1} = z_c, \; Z_c = \sum_{i=1}^{c} c_i z_i.$

Then $c \leq b$. Consider also the linear functions

(11) $\quad z_{c+1} = Z_c, \; z_{c+2} = Z_{c+1}, \ldots, z_{b+1} = Z_b,$

which may be computed in terms of $z_1, \ldots, z_c$ by the recursion formula

(12) $\quad z_{c+j} = \sum_{i=1}^{c} c_i z_{i+j-1} \quad (j = 1, \ldots, b - c + 1).$

The chain with the leader $w_1 = y_1 + z_1$ has the successive terms $w_j = y_j + z_j (j = 1, \ldots, b)$, which are linearly independent since

$y_1, \ldots, y_b, z_1, \ldots, z_c$ are independent. For, if $\sum k_j(y_j + z_j) = 0$, then $\sum k_j y_j + Z = 0$, where Z is a linear function of $z_1, \ldots, z_c$, whence each $k_j = 0$, $Z = 0$. Since b is the maximal length of a chain whose leader is linearly independent of $x_1, \ldots, x_a$, the chain led by w_1 is of length $\leq b$, and hence is of length b. Write $W_j = Y_j + Z_j$. Then

$$W_b = \sum_{i=1}^{b} b_i y_i + z_{b+1} = [w_1, \ldots, w_b]$$

$$= \sum_{i=1}^{b} b_i w_i = \sum_{i=1}^{b} b_i(y_i + z_i),$$

whence

(13) $$z_{b+1} = \sum_{i=1}^{b} b_i z_i.$$

If $c = b$, then $z_{b+1} = \sum c_i z_i$ by (10) or (12), whence $c_i = b_i$, so that (9) and (10) have the same characteristic determinant.

Next, let $c < b$. If from the difference of the two members of (13) we eliminate $z_{b+1}, z_b, \ldots, z_{c+1}$ in turn by means of (12), we obtain a linear function of $z_1, \ldots, z_c$, whose coefficients must all be zero. By a mere change of notation, we conclude that, if we eliminate $\lambda^{b+1}, \lambda^b, \ldots, \lambda^{c+1}$ in turn from

(14) $$\lambda^{b+1} - \sum_{i=1}^{b} b_i \lambda^i$$

by means of

(15) $\quad \lambda^{c+j} = \sum_{i=1}^{c} c_i \lambda^{i+j-1} \qquad (j = 1, \ldots, b - c + 1),$

we obtain a linear function of $\lambda, \ldots, \lambda^c$ whose coefficients are all zero. These eliminations are evidently the successive steps in the ordinary process of division of (14) by

(16) $$\lambda^{c+1} - \sum_{i=1}^{c} c_i \lambda^i.$$

Hence (14) is exactly divisible by (16). Removing the factor λ from each, we obtain the characteristic determinants of (9) and (10) respectively, signs apart. For, if N is the matrix of (9), its λ-matrix $N - \lambda I$ is

(17) $$\begin{bmatrix} -\lambda & 1 & 0 & 0 & \cdots & 0 \\ 0 & -\lambda & 1 & 0 & \cdots & 0 \\ \hdotsfor{6} \\ b_1 & b_2 & b_3 & b_4 & \cdots & b_b - \lambda \end{bmatrix},$$

whose determinant, called the characteristic determinant of N, is

(18) $\quad (-1)^b (\lambda^b - b_1 - b_2 \lambda - b_3 \lambda^2 - \cdots - b_b \lambda^{b-1}).$

EXAMPLE 1. Let S be $W_1 = aw_1 + bw_2$, $W_2 = w_1$, $W_3 = bw_2 + aw_3$. Employing the new variables $x_1 = w_2$, $x_2 = w_1$, $y = w_3 - w_1$, $X_1 = W_2$, etc., we get
$$C: \quad X_1 = x_2, \quad X_2 = bx_1 + ax_2, \quad Y = ay.$$
Apparently we have a canonical form composed of a chain x_1, x_2 and a chain y. But the characteristic determinant $\lambda^2 - a\lambda - b$ of the partial transformation on x_1 and x_2 is divisible by the characteristic determinant $a - \lambda$ of $Y = ay$ only when $b = 0$. If $b = 0$, C is the true canonical form of S, since every linear function of x_1, x_2, y, in which the coefficient of x_1 is not zero, is the leader of a chain of length 2; while, if that coefficient is zero, it is the leader of a chain of length 1. If $b \neq 0$, take
$$z_1 = x_1 + y, \quad Z_1 = X_1 + Y = x_2 + ay \equiv z_2,$$
$$Z_2 = X_2 + aY = bx_1 + ax_2 + a^2 y \equiv z_3.$$
The determinant of the coefficients of x_1, x_2, y is $-b$. A true canonical form when $b \neq 0$ is therefore
$$Z_1 = z_2, \quad Z_2 = z_3, \quad Z_3 = -abz_1 + (b - a^2)z_2 + 2az_3.$$

EXAMPLE 2. There is a single chain, say with the leader $x_1 = w_1$, in the canonical form of
$$W_1 = w_2 + w_3, \quad W_2 = w_1 + w_3, \quad W_3 = w_1 + w_2 + w_3.$$

43. Theorem 3. *If $d_1, \ldots, d_k$ are the characteristic determinants of the partial transformations on the variables of the first, $\ldots$, kth chains of a canonical linear transformation whose λ-matrix M has n rows, and if g_i denotes the greatest common divisor of all $(n-i)$-rowed determinants of M, then*

$$g_1 = d_2 \cdots d_k, \quad g_2 = d_3 \cdots d_k, \ldots, g_{k-1} = d_k, \quad g_j = 1 \, (j \geqq k).$$

We employ the condensed notation

$$M = \begin{bmatrix} A_1 & 0 & 0 & \cdots & 0 & 0 \\ 0 & A_2 & 0 & \cdots & 0 & 0 \\ \multicolumn{6}{c}{\dotfill} \\ 0 & 0 & 0 & \cdots & 0 & A_k \end{bmatrix}.$$

where $A_1, \ldots, A_k$ are the λ-matrices of the partial transformations on the variables of the first, $\ldots$, kth chains of lengths $a, b, \ldots, e$. For example, A_2 is given by (17). Their determinants are $d_1, \ldots, d_k$. Let $a_{ij}, b_{ij}, \ldots$ be the elements of $A_1, A_2, \ldots$ in the ith row and jth column.

The minor M' of an element in one of the 0 matrices of M is zero. For example, let M' be obtained by deleting a column containing elements of A_1 and a row not containing elements of A_1. Evidently every a-rowed minor formed from the first a rows of M' has a column of zero elements. Hence Laplace's development of M' according to the first a rows gives $M' = 0$.

The minor of b_1 in (17) is 1. Similarly, the minor of a_{a1} in A_1 is 1. Hence the minor of a_{a1} in M is $g_1 = d_2 \cdots d_k$, while the minor of any a_{ij} is evidently a multiple of g_1. The minor of any b_{ij} is a multiple of $d_1 d_3 d_4 \cdots d_k$ and hence of g_1, since d_1 is a multiple of d_2 by Theorem 2. The minor of any c_{ij} is a multiple of $d_1 d_2 d_4 \cdots d_k$ and hence of g_1. In this manner we conclude that g_1 is the greatest common divisor of all $(n-1)$-rowed minors of M.

Any $(n-2)$-rowed minor is zero if it is obtained from M by deleting two rows and two columns crossing in elements of a zero matrix; the proof is the same as for M' above. The minor obtained by deleting from M the row and column crossing in a_{a1} as well as those crossing in b_{b1} is evidently $g_2 = d_3 \cdots d_k$, while the minor obtained by deleting the row and column crossing in any a_{ij} as well as those crossing in any b_{ij} is clearly a multiple of g_2. The minor obtained by deleting from M the row and column crossing in any a_{ij} as well as those crossing in any c_{ij} is evidently a multiple of $d_2 d_4 \cdots d_k$ and hence of g_2.

Proceeding in this manner, we have Theorem 3.

44. Invariant factors of a canonical transformation. By (18), we have

(19) $$d_1 = (-1)^a I_n, \quad d_2 = (-1)^b I_{n-1}, \ldots,$$
$$d_k = (-1)^e I_{n-k+1},$$

where $I_n, \ldots, I_{n-k+1}$ are polynomials in λ in which the coefficient of the highest power of λ is unity. Write

(19') $$I_{n-k} = 1, \ldots, I_1 = 1.$$

It is customary to call I_j the jth *invariant factor* of the λ-matrix of the canonical transformation C. Hence, apart from sign, the invariant factors distinct from unity are the characteristic determinants of the partial transformations, one for each chain of C.

THEOREM 4. *If I_j is the jth invariant factor of the n-rowed λ-matrix M of a canonical transformation, then I_j divides I_{j+1}. If G_j is the greatest common divisor of all j-rowed determinants of M chosen so that the coefficient of the highest power of λ in G_j is unity, and if $G_0 = 1$, then*

(20) $\quad\quad G_j = I_1 I_2 \cdots I_j \quad\quad (j = 1, \ldots, n),$

(21) $\quad\quad I_j = G_j/G_{j-1} \quad\quad (j = 1, \ldots, n),$

so that the G's uniquely determine the I's, and conversely.

The first statement follows from Theorem 2. Next, if we write $d_j = 1$ when $j > k$, we see by Theorem 3 that $G_{n-i} = \pm d_{i+1} \cdots d_n$ for $0 \leq i < n$. By (19) and (19'), $d_j = \pm I_{n-j+1}$ for $1 \leq j \leq n$. Hence we have (20) and therefore also (21).

45. Theorem 5. *If in a linear transformation with the matrix A, we introduce new variables defined by a non-singular linear transformation with the matrix B, we obtain one with the matrix BAB^{-1}.*

Using the same notation for a linear transformation and its matrix, let

$$A: \quad X_i = \sum_{j=1}^{n} a_{ij} x_j \qquad (i = 1, \ldots, n),$$

(22) $\quad B: \quad y_i = \sum_{k=1}^{n} b_{ik} x_k \qquad (i = 1, \ldots, n),$

(23) $\quad B: \quad Y_i = \sum_{k=1}^{n} b_{ik} X_k \qquad (i = 1, \ldots, n).$

Since $|b_{ik}| \neq 0$, we may solve equations (22), and obtain

$$B^{-1}: \quad x_j = \sum_{s=1}^{n} c_{js} y_s \qquad (j = 1, \ldots, n).$$

Elimination of the X_k and the x_j between (23), A, and B^{-1} gives

$$BAB^{-1}: \quad Y_i = \sum_{k,\, j,\, s}^{1, \ldots, n} b_{ik} a_{kj} c_{js} y_s \qquad (i = 1, \ldots, n).$$

46. Rotations and orthogonal transformations. We shall illustrate the introduction of new variables by an important example. If w_1, w_2, w_3 are the coordinates of an arbitrary point W referred to given rectangular axes Ow_1, Ow_2, Ow_3, we seek the coordinates

§ 46] ROTATIONS, ORTHOGONAL TRANSFORMATIONS 99

p_1, p_2, p_3 of the point P derived from W by the rotation about ON through angle α counterclockwise when viewed from N toward O. Let n_1, n_2, n_3 be the direction cosines of ON. Choose new rectangular axes Ox, Oy, Oz, whose direction cosines with respect to Ow_1, Ow_2, Ow_3 are l_1, l_2, l_3; m_1, m_2, m_3; n_1, n_2, n_3, respectively. Referred to the new axes, let W and P have the coordinates x, y, z and X, Y, Z, respectively. Since the axis of rotation ON is now Oz, we have

(24) $X = x \cos \alpha - y \sin \alpha, \quad Y = x \sin \alpha + y \cos \alpha, \quad Z = z,$

by plane analytics. By solid analytics, we have

$$
\begin{aligned}
& x = l_1 w_1 + l_2 w_2 + l_3 w_3, & w_1 = l_1 x + m_1 y + n_1 z, \\
(25) \quad & y = m_1 w_1 + m_2 w_2 + m_3 w_3, & w_2 = l_2 x + m_2 y + n_2 z, \\
& z = n_1 w_1 + n_2 w_2 + n_3 w_3, & w_3 = l_3 x + m_3 y + n_3 z.
\end{aligned}
$$

Let D be the determinant of the transformation T defined by the first triple of equations. The second triple defines the inverse transformation T^{-1}, whose determinant is derived from D by interchanging rows and columns. Hence $DD = 1$, $D = \pm 1$. After reversing if necessary the positive direction of one of the new axes, we have $D = +1$. Solving the second triple of equations for z and comparing the result with $z = n_1 w_1 + \ldots$, we see that

(26) $l_2 m_3 - l_3 m_2 = n_1, \quad l_1 m_3 - l_3 m_1 = -n_2, \quad l_1 m_2 - l_2 m_1 = n_3.$

We must express transformation (24) in the new variables w_1, w_2, w_3 defined by (25) and

$$p_i = l_i X + m_i Y + n_i Z \qquad (i = 1, 2, 3).$$

In the latter insert the expressions (24) and then the values of x, y, z from (25). We get

$$p_i = \sum_{j=1}^{3} [(l_i l_j + m_i m_j)\cos \alpha + (l_j m_i - l_i m_j) \sin \alpha + n_i n_j] w_j.$$

By the properties of direction cosines of perpendicular lines, the coefficient of $\cos \alpha$ is equal to $-n_i n_j$ or $1 - n_i^2$, according as $j \neq i$ or $j = i$. The coefficient of $\sin \alpha$ is one of the values (26) or zero in the corresponding cases. Write

$$(27) \qquad d = \cos \tfrac{1}{2}\alpha, \qquad a_i = n_i \sin \tfrac{1}{2}\alpha \qquad (i = 1, 2, 3).$$

Then $\cos \alpha = 2d^2 - 1$. Thus the coefficient of w_i in p_i is

$$(1 - n_i^2) \cos \alpha + n_i^2 = 2d^2 - 1 + 2(n_i \sin \tfrac{1}{2}\alpha)^2.$$

Next, for $j \neq i$, the coefficient of w_j is

$$n_i n_j (1 - \cos \alpha) + e_{ji} n_k \sin \alpha = 2 n_i n_j \sin^2 \tfrac{1}{2}\alpha \\ + 2 e_{ji} n_k \sin \tfrac{1}{2}\alpha \cos \tfrac{1}{2}\alpha,$$

where k denotes that one of 1, 2, 3 which is distinct from i and j, while

$$e_{12} = e_{23} = e_{31} = +1, \qquad e_{13} = e_{21} = e_{32} = -1.$$

Inserting the values from (27), we get

$$p_i = (2d^2 - 1 + 2a_i^2) w_i + \sum_{j \neq i}^{1,2,3} [2a_i a_j + 2 e_{ji} a_k d] w_j,$$

or, written at length,

$$(28) \quad \begin{aligned} p_1 &= (2a_1^2 + 2d^2 - 1) w_1 + 2(a_1 a_2 - a_3 d) w_2 + 2(a_1 a_3 + a_2 d) w_3, \\ p_2 &= 2(a_1 a_2 + a_3 d) w_1 + (2 a_2^2 + 2d^2 - 1) w_2 + 2(a_2 a_3 - a_1 d) w_3, \\ p_3 &= 2(a_1 a_3 - a_2 d) w_1 + 2(a_2 a_3 + a_1 d) w_2 + (2 a_3^2 + 2d^2 - 1) w_3. \end{aligned}$$

Theorem 6. *Formula (28) represents the rotation about the axis ON with the direction cosines n_1, n_2, n_3 through angle α counterclockwise when viewed from N toward O.*

§ 46] ROTATIONS, ORTHOGONAL TRANSFORMATIONS 101

The values of d, a_1, a_2, a_3 are given by (27), whence $a_1^2 + a_2^2 + a_3^2 + d^2 = 1$.

Consider a real ternary orthogonal transformation

$$T: \quad y_i = b_{i1} x_1 + b_{i2} x_2 + b_{i3} x_3 \quad (i = 1, 2, 3),$$

whose matrix B is of determinant $+1$. Thus $B'B = I$. If D is the determinant of $B - I$ and hence of $B' - I$, then

$$B'(B - I) = -(B' - I), \quad |B'|D = -D, \quad 2D = 0.$$

The point $N = (n_1, n_2, n_3)$ will be unaltered by T if

$$n_i = b_{i1} n_1 + b_{i2} n_2 + b_{i3} n_3 \quad (i = 1, 2, 3),$$

which may be written as homogeneous equations having the determinant $D = 0$. Hence there exist real solutions n_1, n_2, n_3, not all zero. Let T replace the point $X = (x_1, x_2, x_3)$ by $Y = (y_1, y_2, y_3)$. Then

$$y_i - n_i = b_{i1}(x_1 - n_1) + b_{i2}(x_2 - n_2) + b_{i3}(x_3 - n_3).$$

Hence $\sum(y_i - n_i)^2 = \sum(x_i - n_i)^2$, for the same reason that $\sum y_i^2 = \sum x_i^2$ for the orthogonal T. Hence if O is the origin, $OY = OX$, $NY = NX$. Thus the triangles OXN and OYN are congruent. Hence X and Y lie on a circle whose plane is perpendicular to ON, so that T represents a rotation about ON.

Any orthogonal transformation of determinant -1 is evidently the product of one of determinant $+1$ by $X = x, Y = y, Z = -z$. The latter represents a reflexion in the xy-plane.

THEOREM 7. *Every real ternary orthogonal transformation represents a rotation or a rotation followed by a reflexion, according as its determinant is $+1$ or -1.*

COROLLARY. *Every real ternary orthogonal transformation of determinant $+1$ is represented by formula* (28).

Cayley[1] expressed the n^2 coefficients of the n-ary orthogonal transformation in terms of $\frac{1}{2}n(n-1)$ parameters; but his formulas do not include all orthogonal transformations, except as limiting cases. See Exs. 3–6 of §54. This subject is only a special topic of the theory[2] of linear transformations which leave unaltered a quadratic or bilinear form.

47. Theorem 8. *Let A and B be n-rowed square matrices such that the elements of A are polynomials in λ, while those of B are independent of λ, and $|B| \neq 0$. Then the greatest common divisor $D_t(\lambda)$, with leading coefficient unity, of all t-rowed determinants of AB is equal to that of A.*

Let d_t be the greatest common divisor of all t-rowed determinants of A. By Theorem 5 of Ch. III, D_t is divisible by d_t. Since A is the product of AB by B^{-1}, the same theorem shows that d_t is divisible by D_t. The two results imply $D_t = d_t$.

Since a like theorem holds for BA, we deduce the

COROLLARY. *Let A, P, Q be n-rowed square matrices such that the elements of A are polynomials in λ, while those of P and Q are independent of λ, and P and Q are non-singular. Then the greatest common divisor of all t-rowed determinants of PAQ is equal to that of A.*

48. In particular, let A be the λ-matrix of any linear transformation T, the matrix of whose coefficients may also be denoted by T, so that $A = T - \lambda I$. By the introduction of new variables defined by a linear transformation with the non-singular matrix B, T becomes a transformation C, the matrix of whose coefficients

---

[1] Scott's Theory of Determinants, revised by Mathews, 1904, 197–202; Pascal's Determinants, §§47–50, which cites twenty writers on orthogonal transformations.

[2] Encyclopédie des sc. math., t. I, v. II, 478–520. Cf. Muth, Elementartheiler, 1899, 160–179. For Hermitian forms, Loewy, Nova Acta Leop. Carol. Akad., 71, 1898, 379–416; extract in Math. Annalen, 50, 1898, 557–76.

may also be designated by C. By Theorem 5, $C = BTB^{-1}$. Write $D = C - \lambda I$. Then $D = BAB^{-1}$. In view of the Corollary, the greatest common divisor of all t-rowed determinants of D is equal to that of A. This result may be stated as follows:

THEOREM 9. *Let any linear transformation T become C by the introduction of any new independent variables. Then the greatest common divisor G_t of all t-rowed determinants of the λ-matrix of T is equal to that of C.*

49. Canonical form determined by invariant factors. Let the new variables be chosen so that C is a canonical form of T. Since the λ-matrices of these transformations have the same G_t, they have the same $I_j = G_j/G_{j-1}$ which, by (19), uniquely determine the characteristic determinants of the partial transformations on the variables of the various chains of C, and hence, by (17) and (18), uniquely determine the matrix of C.

THEOREM 10. *For any linear transformation T, the quotient $I_j = G_j/G_{j-1}$ is a polynomial in λ called the jth invariant factor of the λ-matrix of T. The matrix of the canonical form of T is uniquely determined by these invariant factors.*

In other words, any linear transformation has a single canonical form apart from the notation for the variables.

50. Consider the transformation T defined by (1), (2), (3), etc., such that the characteristic determinant of (1) is divisible by that of (2), while the latter is divisible by that of (3), etc. Then T is its own canonical form, so that $a, b, c, \ldots$ have the properties stated at the end of Theorem 1. Hence we have

THEOREM 11. *There exists a linear transformation with coefficients in any field F whose λ-matrix has any prescribed invariant factors I_j with coefficients in F such that I_j divides I_{j+1} for every j.*

51. Similar transformations. Two linear transformations (or matrices) S and T in F (i.e., with coefficients in F) are called *similar in F* if there exists a non-singular transformation B in F such that $BSB^{-1} = T$. In other words, S becomes T by the introduction of new variables defined by transformation B. Theorem 9 states that the λ-matrices of two similar transformations have the same G_t.

Conversely, if the λ-matrices of S and T have the same G_t, then S and T are similar in F. Let $U = QSQ^{-1}$ and $V = RTR^{-1}$ be the canonical forms of S and T. Since the canonical matrices U and V have the same G_t, they are identical by Theorem 10. Hence

$$QSQ^{-1} = V, \quad R^{-1}VR = T, \quad (R^{-1}Q)S(R^{-1}Q)^{-1} = T,$$

so that S and T are similar.

THEOREM 12. *Two linear transformations (or matrices) in F are similar in F if and only if their λ-matrices have the same invariant factors.*

52. Invariant factors. Let M and N be any two n-rowed square matrices of which N is non-singular. Write $A = MN^{-1}$. Then $M - \lambda N = (A - \lambda I)N$ has the same G_j as $A - \lambda I$ by Theorem 8. For the latter, $I_j = G_j/G_{j-1}$ is a polynomial in λ by Theorem 10, and hence may be defined as the jth invariant factor of $M - \lambda N$.

Removing the restriction that N be non-singular, let G_j denote the greatest common divisor of all j-rowed determinants of $\rho M + \sigma N$, where ρ and σ are independent variables. Since each such determinant may be expanded according to the elements of a row (or column) and hence is a linear combination of certain $(j-1)$-rowed determinants of $\rho M + \sigma N$, G_{j-1} is a factor of every j-rowed determinant and hence of G_j. The quotient G_j/G_{j-1} is therefore a homogeneous polynomial in ρ and σ, and is called the jth invariant factor of $\rho M + \sigma N$.

Exercises

1. $X_1 = x_2$, $X_2 = ax_2$, $X_3 = ax_3$ has the invariant factors $\lambda(\lambda - a)$ and $\lambda - a$.

2. The invariant factors of
$$\begin{bmatrix} 0 & 0 & 2-\lambda \\ 0 & 1+\lambda & 2\lambda \\ 2-\lambda & 0 & 0 \end{bmatrix}$$
are $(\lambda - 2)(1 + \lambda)$ and $\lambda - 2$.

53. Classic canonical form.[1] For certain purposes a canonical form involving irrationalities is preferable to that in Theorem 1.

THEOREM 13. *Start with the canonical form in Theorem 1, with coefficients in a field F, composed of partial transformations of the type*

$$(29) \quad X_1 = x_2, \quad X_2 = x_3, \ldots, X_{a-1} = x_a, \quad X_a = \sum_{i=1}^{a} a_i x_i.$$

Let the characteristic equation $\Delta(\lambda) = 0$ of (29) have the roots $R, S, T, \ldots$, of multiplicities $r, s, t, \ldots$, respectively. By the introduction of new variables $u_i, v_i, w_i, \ldots$, which are linearly independent homogeneous linear functions of $x_1, \ldots, x_a$, (29) can be reduced to the classic canonical form

$$(30) \quad \begin{cases} U_1 = Ru_1, & U_j = Ru_j + u_{j-1} & (j = 2, \ldots, r), \\ V_1 = Sv_1, & V_j = Sv_j + v_{j-1} & (j = 2, \ldots, s), \\ W_1 = Tw_1, & W_j = Tw_j + w_{j-1} & (j = 2, \ldots, t), \ldots \end{cases}$$

Here the u_i (or v_i, w_i) are homogeneous linear functions of the x's whose coefficients are polynomials in R (or S, T) with coefficients in F. If R and S are roots of the same equation irreducible in F, v_i may be derived from u_i by replacing R by S.

---

[1] Jordan, Traité des Substitutions, 1870, 114–26, for transformations with integral coefficients taken modulo p of determinant prime to p. For any field, Dickson, Amer. Jour. Math., 24, 1902, 101–8.

Before giving a proof leading to explicit formulas for the $u_i, v_i, \ldots$, which imply (30) and their further properties, we shall verify *a fortiori* that (29) is similar to (30), when F is the field of all complex numbers, and hence can be reduced to (30) by the introduction of new variables $u_i, v_i, \ldots$, formulas for which are, however, not found by this method.

The λ-matrix of the transformation in the first line of (30) is

$$A_1 = \begin{bmatrix} R-\lambda & 0 & 0 & \cdots & 0 & 0 & 0 \\ 1 & R-\lambda & 0 & \cdots & 0 & 0 & 0 \\ \multicolumn{7}{c}{\dotfill} \\ 0 & 0 & 0 & \cdots & 0 & 1 & R-\lambda \end{bmatrix}.$$

Let A_2 and A_3 be the λ-matrices of the transformations in the second and third lines of (30). Thus the λ-matrix of (30) is

$$M = \begin{bmatrix} A_1 & 0 & 0 & 0 & \cdots \\ 0 & A_2 & 0 & 0 & \cdots \\ 0 & 0 & A_3 & 0 & \cdots \\ \multicolumn{5}{c}{\dotfill} \end{bmatrix}.$$

The determinant of A_1 is $d_1 = (R-\lambda)^r$, while the minor of the last element of the first row of A_1 is unity. The determinants of $A_2, A_3, \ldots$ are $d_2 = (S-\lambda)^s, d_3 = (T-\lambda)^t, \ldots$. Then the determinant of M is $\Delta = d_1 d_2 d_3 \ldots$, while the first minors of M include $d_2 d_3 d_4 \ldots, d_1 d_3 d_4 \ldots, d_1 d_2 d_4 \ldots$, etc. Their greatest common divisor is 1, since it has none of the factors $R-\lambda$, $S-\lambda, \ldots$. Again, the greatest common divisor of the first minors of the λ-matrix of (29) is 1, since the minor of b_1 in (17) is 1. Since the λ-matrices of (29) and (30) therefore have the same invariant factors $1, \ldots, 1, \pm\Delta$, they are similar by Theorem 12. In other words, there exist new variables $u_i, v_i, \ldots$ whose introduction reduces (29) to (30).

The properties of the $u_i, v_i, \ldots$ stated at the end of Theorem 13 are important in various applications, but are not used in the present text. Some readers may prefer therefore to omit the following proof of those properties in connection with a direct deduction of (30).

We again start with (29), in which the a_i belong to the field F. As in (18), the characteristic determinant of (29) is the product of $(-1)^a$ by $C_a(\lambda)$, where

(31) $\quad C_{a-k}(\lambda) \equiv \lambda^{a-k} - \sum_{i=0}^{a-k-1} a_{i+k+1}\lambda^i \quad (k = 0, 1, \ldots, a-1).$

Write $C_0(\lambda) = 1$ as the definition of (31) when $k = a$. Then (29) replaces the function

$$f(\lambda) \equiv \sum_{k=1}^{a} x_k\, C_{a-k}(\lambda)$$

by

$$Q(\lambda) \equiv \sum_{j=2}^{a} x_j\, Q_j + a_1 x_1,$$

where

$$Q_j \equiv C_{a-j+1}(\lambda) + a_j = \lambda^{a-j+1} - \sum_{i=0}^{a-j} a_{i+j}\lambda^i + a_j = \lambda C_{a-j}(\lambda).$$

Hence

(32) $\quad Q(\lambda) - \lambda f(\lambda) \equiv x_1[a_1 - \lambda C_{a-1}(\lambda)] \equiv -x_1 C_a(\lambda),$

where the last reduction follows from the preceding result for $j = 1$. By t differentiations with respect to λ, we get

(33) $\quad Q^{(t)}(\lambda) - tf^{(t-1)}(\lambda) - \lambda f^{(t)}(\lambda) \equiv -x_1 C_a^{(t)}(\lambda).$

Let R be a root of $C_a(\lambda) = 0$ of multiplicity r, and write u_1 for $f(R)$, and u_{t+1} for the quotient[1] of $f^{(t)}(R)$ by $t!$. Since the trans-

---

[1] The division is exact for each coefficient. The entire proof is valid also when F is a so-called modular field in which the result of adding any element to itself p times is the element zero. An example is the field of the p residues of integers modulo p.

formation (29) replaces $f(\lambda)$ by $Q(\lambda)$ and hence replaces $f^{(t)}(\lambda)$ by $Q^{(t)}(\lambda)$ for every λ, it replaces u_{t+1} by $Q^{(t)}(R) \div t!$, which we therefore denote by U_{t+1}. Taking $\lambda = R$ in (32) and (33), we get

$$U_1 = Ru_1, \quad U_{t+1} = Ru_{t+1} + u_t \quad (t = 1, \ldots, r-1).$$

Writing j for $t + 1$, we obtain the first line of (30). By their origin, the u_i are linear functions of $x_1, \ldots, x_a$ whose coefficients are polynomials in R with coefficients in F.

It remains only to prove that the new variables are linearly independent. If

$$(34) \qquad \sum_{j=1}^{r} g_j u_j + \sum_{j=1}^{s} h_j v_j + \sum_{j=1}^{t} k_j w_j \equiv 0,$$

identically in $x_1, \ldots, x_a$, we are to prove that the constants g, h, k are all zero. From the corresponding identity in the U, V, W, we insert the values (30), subtract the product of (34) by R, and get

$$(35) \quad \sum_{j=2}^{r} g_j u_{j-1} + \sum_{j=1}^{s} h_j (S-R) v_j + \sum_{j=2}^{s} h_j v_{j-1}$$
$$+ \sum_{j=1}^{t} k_j (T-R) w_j + \sum_{j=2}^{t} k_j w_{j-1} \equiv 0.$$

If we can prove that the coefficients of u_i, v_i, w_i in (35) are all zero, we shall have $k_t = 0$ by w_t, then $k_{t-1} = 0$ by $w_{t-1}, \ldots$, $k_1 = 0$ by w_1, and similarly every $h_i = 0$ by v_i and every $g_j = 0$ ($j > 1$), whence (34) becomes $g_1 u_1 = 0$. But the coefficient of x_a in $u_1 = f(R)$ is $C_0(R) = 1$. Hence also $g_1 = 0$, so that all coefficients in (34) would be zero.

Since (35) is of the form (34), but lacks u_r, it suffices to prove that a relation of type (34) which lacks u_r has every coefficient zero. The new identity (35) now lacks both u_r and u_{r-1}. Taking it in place of (34), we see by repetitions of this argument that the problem reduces to one involving only the v_i and w_i. In the corresponding identity in V_i and W_i, we insert the values (30), and subtract the product of the former identity by S. Then as before we eventually obtain a relation involving only the w_i. This time we subtract the product by T.

54. Elementary divisors. In the classic canonical form (30) of (29), we saw that the characteristic determinants of the partial transformations, given by the successive lines of (30), are, apart from sign,

$$(36) \qquad (\lambda - R)^r, \quad (\lambda - S)^s, \quad (\lambda - T)^t, \ldots,$$

whose product, apart from sign, is the characteristic determinant $\Delta(\lambda)$ of (29), i.e., its unique invariant factor distinct from 1. These divisors (36) are called the *elementary divisors* of $\Delta(\lambda)$ or of the λ-matrix of (29). Recalling the definition (19) of the invariant factors of the λ-matrix of any canonical transformation, of which our (29) is only one partial transformation, we obtain

THEOREM 14. *The elementary divisors of the λ-matrix of any linear transformation are obtained from its invariant factors other than 1 by taking the highest power of each linear factor of each such invariant factor which is a divisor of the latter. Conversely, from a list of the elementary divisors written in any order, we may determine uniquely the invariant factors I_j other than 1 by using the fact that I_j divides I_{j+1}.*

For example, let the invariant factors be

$$I_1 = 1, \qquad I_2 = \lambda, \qquad I_3 = \lambda(\lambda - 1)^2, \qquad I_4 = \lambda^2(\lambda - 1)^3.$$

Then the elementary divisors are

$$\lambda^2, \lambda, \lambda, (\lambda - 1)^3, (\lambda - 1)^2.$$

Conversely, given the latter, we deduce I_4 as the product of the highest powers of the distinct linear functions, then I_3 as the product of the next highest powers, etc.

The classic canonical form shows that there exists a linear transformation on n variables whose λ-matrix has any preassigned elementary divisors the sum of whose degrees is n.

Our definition of elementary divisors applies only to λ-matrices $M - \lambda I$. In §52 we defined the invariant factors I_j of $P = \rho M + \sigma N$, where M and N are any n-rowed square matrices. All the I_j distinct from 1 may be expressed as products of powers of the same linear functions of ρ and σ, no two of which have proportional coefficients. Then those powers which have exponents > 0 are called the elementary divisors of P.

EXERCISES

1. The three types of canonical linear transformations on three variables are

$$U = Ru, \quad V = Sv, \quad W = Tw;$$
$$U_1 = Ru_1, \quad U_2 = Ru_2 + u_1, \quad V = Sv;$$
$$U_1 = Ru_1, \quad U_2 = Ru_2 + u_1, \quad U_3 = Ru_3 + u_2.$$

2. Find the elementary divisors of the λ-matrix of each transformation in Ex. 1.

3. Verify that formula (28) for any rotation in space is equivalent to the equation $p = qwq'$ in real quaternions

$$q = dI + a_1 i + a_2 j + a_3 k, \quad w = w_1 i + w_2 j + w_3 k,$$
$$p = p_1 i + p_2 j + p_3 k,$$

where q is of norm unity. See Ex. 6.

4. The equation $\xi = xy$ in matrices

$$\xi = \begin{pmatrix} \xi_1 & \xi_2 \\ \xi_3 & \xi_4 \end{pmatrix}, \quad x = \begin{pmatrix} x_1 & x_2 \\ x_3 & x_4 \end{pmatrix}, \quad y = \begin{pmatrix} y_1 & y_2 \\ y_3 & y_4 \end{pmatrix}$$

defines a homogeneous linear transformation on $x_1, \ldots, x_4$ which evidently leaves invariant the quadric surface S defined by $|x| = 0$; verify that it leaves invariant each of the lines

$$L_k : \quad x_1 = kx_3, \quad x_2 = kx_4$$

on S. Similarly, $\xi = yx$ leaves invariant each line

$$l_k : \quad x_1 = kx_2, \quad x_3 = kx_4,$$

also on S. Then[1] $\xi = zxy$ leaves S invariant, and permutes the lines L_k amongst themselves and likewise the l_k.

5. In Ex. 4 introduce new variables X_r and new parameters Y_r such that S becomes $\Sigma X_r{}^2 = 0$:

$$x_1 = X_1 + iX_4, \ x_4 = X_1 - iX_4, \ x_2 = -X_2 + iX_3, \ x_3 = X_2 + iX_3,$$

$$y_1 = Y_4 - iY_1, \ y_4 = Y_4 + iY_1, \ y_2 = Y_3 + iY_2, \quad y_3 = -Y_3 + iY_2.$$

Then to $\xi = xy$ corresponds $\Xi = XY$ in quaternions $X = X_1 i + X_2 j + X_3 k + X_4 I$. Thus if Z and Y are real quaternions, $\Xi = ZXY$ gives every real linear transformation of positive determinant which leaves $\Sigma X_r{}^2$ invariant up to a constant factor.[2]

6. Deduce from Ex. 5 the real orthogonal transformations on three variables. Take $X_4 = 0$, $\Xi_4 = 0$. In other words, take $X + X' = 0$, $\Xi + \Xi' = 0$. Hence shall $ZXY = Y'XZ'$. Write p for $Z'Y^{-1}$. Then $p'X = Xp$. The case $X = i$ shows that the coefficient of i in p is zero. By symmetry, the cases $X = j$ and $X = k$ show that the coefficients of j and k in p are zero. Hence $p = rI$, where r is a real number. Thus $Z' = rY$ and $\Xi = ZXZ' r^{-1}$. Taking norms, we see that the condition that $X_1{}^2 + X_2{}^2 + X_3{}^2$ be invariant is $n(Z) = \pm r$, which is therefore positive. Write q for the real quaternion $Z(\pm r)^{-\frac{1}{2}}$ of norm unity. Thus $\Xi = \pm qXq'$. The sign is in fact $+$ or $-$, according as the determinant of the orthogonal transformation is $+1$ or -1. This gives a new proof of Ex. 3.

---

[1] These transformations and their products by $(x_2\ x_3)$, which interchanges the two sets of lines, are the only linear transformations leaving S invariant. Dickson, Bull. Amer. Math. Soc., 22, 1915, 53–61.

[2] For an elegant proof using only quaternions, see Hurwitz, Zahlentheorie der Quaternionen, Berlin, 1919, 62–66.

Chapter VI

PAIRS OF BILINEAR, QUADRATIC, AND HERMITIAN FORMS

55. Historical note. For a pair of bilinear or quadratic forms (the second being non-singular), the theory was first obtained by Weierstrass[1] by developments in series, which introduce certain square roots. There was a serious gap in his proof. For pairs of bilinear forms, all irrationalities are avoided by the method of Frobenius.[2] However, the resulting rational criterion for the equivalence of pairs of bilinear forms in any field may be proved[3] by simple modifications of Weierstrass's method. We shall give here a still simpler rational method based on the theory of linear transformations.

56. Equivalence of two pairs of matrices. Two pairs of $n \times n$ matrices M, N and U, V, with elements in any field F, are called *equivalent in F* if and only if there exist non-singular n-rowed matrices P and Q with elements in F such that

(1) $$PMQ = U, \quad PNQ = V.$$

We shall first treat the case in which N and V are non-singular (cf. §57). Suppose that two such pairs are equivalent. Then $MN^{-1} = J$ is similar to $UV^{-1} = W$ since

$$PJP^{-1} = PMQ \cdot Q^{-1} N^{-1} P^{-1} = UV^{-1} = W,$$

[1] Berlin. Berichte, 1868, 310; Werke, II, 19.

[2] Jour. für Math., 86, 1879, 146–208 (see p. 202). For a report on this and related papers, see the author's History of the Theory of Numbers, III, 1923, 284–8.

[3] Dickson, Trans. Amer. Math. Soc., 10, 1909, 347–60. With the same modifications, the method was extended to Hermitian forms by M. I. Logsdon, Amer. Jour. Math., 44, 1922, 254–60.

§ 56] EQUIVALENT PAIRS OF MATRICES 113

so that the λ-matrices of J and W have the same invariant factors. By §52, the λ-matrix $J - \lambda I$ of J has the same invariant factors as

$$(J - \lambda I)N = (MN^{-1} - \lambda I)N = M - \lambda N.$$

Likewise, the λ-matrix $W - \lambda I$ of W has the same invariant factors as $U - \lambda V$. Hence $M - \lambda N$ and $U - \lambda V$ have the same invariant factors.

Conversely, the last result implies that the λ-matrices of J and W have the same invariant factors, whence J and W are similar in F. In other words, there exists a non-singular matrix P in F such that $PJP^{-1} = W$. Then

$$P(J - \lambda I)P^{-1} = W - \lambda I,$$

$$P(MN^{-1} - \lambda I)P^{-1} = UV^{-1} - \lambda I,$$

$$P(M - \lambda N)N^{-1}P^{-1} = (U - \lambda V)V^{-1}.$$

Writing Q for $N^{-1}P^{-1}V$, we get

$$P(M - \lambda N)Q = U - \lambda V.$$

By the terms free of λ and those containing λ, we get (1).

THEOREM 1. *Two pairs of $n \times n$ matrices M, N and U, V, with elements in any field F, such that N and V are non-singular, are equivalent in F if and only if $M - \lambda N$ and $U - \lambda V$ have the same invariant factors, or, if we prefer, the same elementary divisors.*

We evidently have the same criterion for the equivalence in F of two pairs of bilinear forms with the matrices M, N and U, V.

57. Next, we omit the assumption that the determinant $|N|$ of N is not zero, but assume that $|M + xN|$ is not zero identically in x. Thus we can select a number $v \neq 0$ of F such that $B = M$

$+ vN$ is not singular. Then if the pairs M, N and U, V are equivalent in F, also the pairs M, B and $U, C = U + vV$ are equivalent in F, and C is not singular, and conversely. Hence by Theorem 1, the pairs M, N and U, V are equivalent in F if and only if $M - \lambda B$ and $U - \lambda C$ have the same invariant factors or the same elementary divisors. The latter property holds if and only if the matrices $\rho M + \sigma N$ and $\rho U + \sigma V$ have the same elementary divisors, as will be proved by means of the following

LEMMA. *If* $A = pM + qN, B = uM + vN$, *where* $pv - qu \neq 0$, *so that* $\mu A + \tau B = \rho M + \sigma N$ *for*

(2) $$\rho = p\mu + u\tau, \qquad \sigma = q\mu + v\tau,$$

then the elementary divisors of $\mu A + \tau B$ *are derived from those of* $\rho M + \sigma N$ *by the transformation* (2).

Let $m\rho + n\sigma$ be a factor of all j-rowed determinants of $\rho M + \sigma N$ and let e be the exponent of the highest power of $m\rho + n\sigma$ which divides all those determinants. The transformation (2) replaces any element of $\rho M + \sigma N$ by the corresponding element of $\mu A + \tau B$, and hence replaces any j-rowed determinant of $\rho M + \sigma N$ by one of $\mu A + \tau B$. Hence if (2) replaces $m\rho + n\sigma$ by $a\mu + b\tau$, $(a\mu + b\tau)^e$ is a factor of all j-rowed determinants of $\mu A + \tau B$. It is, moreover, the highest power of $a\mu + b\tau$ which divides those determinants. For if they were all divisible by $(a\mu + b\tau)^g$, where $g > e$, we see by applying the inverse of transformation (2) that $(m\rho + n\sigma)^g$ would be a factor of all j-rowed determinants of $\rho M + \sigma N$, contrary to hypothesis. The Lemma now follows from the definition of elementary divisors in §54.

We may employ the Lemma when $p = u = 1, q = 0, v \neq 0$. Then $A = M, B = M + vN$. Hence the elementary divisors of $\mu M + \tau B$ are derived from those of $\rho M + \sigma N$ by the transformation $\rho = \mu + \tau, \sigma = v\tau$. Thus, if $C = U + vV$, then $\mu M + \tau B$ and $\mu U + \tau C$ have the same elementary divisors if and only if $\rho M + \sigma N$ and $\rho U + \sigma V$ have the same elementary divisors.

Taking $\mu = 1$, $\tau = -\lambda$, we have the notations employed in the discussion preceding the Lemma. This proves

THEOREM 2. *Let M, N, U, V be any n-rowed square matrices with elements in any field F such that neither of the determinants $|\rho M + \sigma N|$ and $|\rho U + \sigma V|$ is zero identically in ρ and σ. Then the pair M, N is equivalent in F to the pair U, V if and only if $\rho M + \sigma N$ and $\rho U + \sigma V$ have the same invariant factors (or the same elementary divisors).*

For brevity we shall speak of the invariant factors of $\rho M + \sigma N$ as the invariant factors of the pair of bilinear forms having the matrices M and N, and speak of the case in which $|\rho M + \sigma N|$ is not zero identically in ρ and σ as *the non-singular case*. Then Theorem 2 may be stated in the following briefer form:

THEOREM 3. *In the non-singular case, two pairs of bilinear forms in $x_1, \ldots, x_n, y_1, \ldots, y_n$ with coefficients in any field F are equivalent under non-singular transformations in F if and only if they have the same invariant factors.*

58. Canonical forms of a pair of bilinear forms. Let their matrices be M and N, where N is assumed for the present to be nonsingular. The pair M, N is equivalent to the pair $J = MN^{-1}$, $NN^{-1} = I$. If P and Q are non-singular and such that $PJQ = W$, $PIQ = I$, then $Q = P^{-1}$. Hence the pair J, I is equivalent to the pair W, I if and only if $PJP^{-1} = W$, i.e., if J and W are similar matrices.

When W is the matrix of the canonical form of the linear transformation of matrix $J = MN^{-1}$, we shall call W, I the canonical pair of M, N.

When all the elements of M and N belong to a field F, we may choose W to be the matrix of the canonical form with coefficients in F of Theorem 1 of Ch. V. This proves

THEOREM 4. *Any pair of bilinear forms in $x_1, \ldots, x_n, y_1, \ldots, y_n$ with coefficients in any field F, such that the second form is non-*

singular, can be reduced by a non-singular linear transformation on the x's and one on the y's, each with coefficients in F, to a unique canonical pair

$$f = \Big(\sum_{i=1}^{a-1} x_i y_{i+1} + x_a \sum_{i=1}^{a} a_i y_i\Big) + \Big(\sum_{i=1}^{b-1} z_i w_{i+1} + z_b \sum_{i=1}^{b} b_i w_i\Big) + \ldots,$$

$$g = \sum_{i=1}^{a} x_i y_i + \sum_{i=1}^{b} z_i w_i + \ldots,$$

where $a + b + \cdots = n$. If there are k of the numbers $a, b, \ldots$, the invariant factors of $f - \lambda g$ are $1, \ldots, 1, I_{n-k+1}, \ldots$,

$$I_{n-1} = \lambda^b - \sum_{i=1}^{b} b_i \lambda^{i-1}, \quad I_n = \lambda^a - \sum_{i=1}^{a} a_i \lambda^{i-1}.$$

Since the $a_i, b_i, \ldots$ are any numbers of F such that I_j divides I_{j+1} for every $j < n$, we have the

COROLLARY. *There exists a pair of bilinear forms f, g in $n + n$ variables with coefficients in F such that $f - \lambda g$ has any preassigned invariant factors I_j with coefficients in F such that I_j divides I_{j+1}.*

Next, if we choose W to be the matrix of the classic canonical form of Theorem 13 of Ch. V, we obtain

THEOREM 5. *Within the field of all complex numbers, any pair of bilinear forms in $n + n$ variables, of which the second form is non-singular, can be reduced to the classic canonical pair*

$$\phi = \Big(R \sum_{i=1}^{r} \xi_i x_i + \sum_{i=2}^{r} \xi_i x_{i-1}\Big) + \Big(S \sum_{i=1}^{s} \eta_i y_i + \sum_{i=2}^{s} \eta_i y_{i-1}\Big) + \ldots,$$

$$\psi = \sum_{i=1}^{r} \xi_i x_i + \sum_{i=1}^{s} \eta_i y_i + \ldots,$$

where $r + s + \cdots = n$. The elementary divisors of $\phi - \lambda\psi$ are $(\lambda - R)^r$, $(\lambda - S)^s$, ...

COROLLARY. There exists a pair of bilinear forms ϕ and ψ in $n + n$ variables, of which ψ is non-singular, such that $\phi - \lambda\psi$ has any preassigned elementary divisors the sum of whose degrees is n.

We now omit the assumption that $|N| \neq 0$, but assume that $|\rho M + \sigma N|$ is not zero identically in ρ, σ. Choose u and v in F so that $B = uM + vN$ is non-singular. Then choose p and q in F so that $pv - qu = 1$, and write $A = pM + qN$. Then

(3) $M = vA - qB, \quad N = -uA + pB.$

The pair of bilinear forms with the matrices A and B can be reduced rationally to a pair f and g of Theorem 4. The same transformations reduce the pair with the matrices M and N to the pair $c = vf - qg$ and $d = -uf + pg$. We have $\rho c + \sigma d = \mu f + \tau g$ for

(4) $\mu = v\rho - u\sigma, \quad \tau = -q\rho + p\sigma,$

which give the solved form of (2). The invariant factors of $\mu f + \tau g$ reduce to those of $f - \lambda g$ by writing $\mu = 1$, $\tau = -\lambda$. Conversely, if we make the invariant factors of $f - \lambda g$ homogeneous in λ, μ, and then replace λ by $-\tau$, we get the invariant factors of $\mu f + \tau g$. The transformation (4) replaces the latter by those of $\rho c + \sigma d$. This proves

THEOREM 6. *In the non-singular case, any pair of bilinear forms in the variables $x_1, \ldots, x_n, y_1, \ldots, y_n$, with coefficients in any field F, can be reduced by a non-singular linear transformation on the x's and one on the y's, each with coefficients in F, to a canonical pair $c = vf - qg$ and $d = -uf + pg$, where v, q, u, p are numbers in F such that $pv - qu = 1$, while f and g are the forms in Theorem 4.*

The invariant factors of $\rho c + \sigma d$ are obtained by applying transformation (4) to those of $\mu f + \tau g$, viz.,

$$1, \ldots, 1, I_{n-k+1}, \ldots,$$

$$I_{n-1} = \tau^b - \sum_{j=0}^{b-1} b_{j+1}\, \tau^j (-\mu)^{b-j},$$

$$I_n = \tau^a - \sum_{j=0}^{a-1} a_{j+1}\, \tau^j (-\mu)^{a-j}.$$

If $(\lambda - R)^r$ is an elementary divisor of $A - \lambda B$, then $(\tau + \mu R)^r$ is one of $\mu A + \tau B$, and conversely. By the Lemma in §57, the inverse (4) of transformation (2) replaces an elementary divisor $(\tau + \mu R)^r$ of $\mu A + \tau B$ by an elementary divisor $(m_1 \rho + n_1 \sigma)^r$ of $\rho M + \sigma N$. Since only the ratio of m_1 and n_1 is material, we may take $m_1 u + n_1 v = 1$. Then if $R = m_1 p + n_1 q$, we have $m_1 \rho + n_1 \sigma = \tau + \mu R$. By Theorem 5 we can reduce the pair of bilinear forms with the matrices A and B to the canonical pair ϕ and ψ. Then that with the matrices (3) is reduced to the pair $\alpha = v\phi - q\psi$ and $\beta = -u\phi + p\psi$, whose expressions reduce to those given in Theorem 7. For, $pv - qu = 1$, $m_1 u + n_1 v = 1$, $m_1 p + n_1 q = R$ imply $m_1 = vR - q$, $n_1 = p - uR$; and, when we employ S instead of R, we replace the subscripts 1 by 2 and hence have $m_2 = vS - q$, $n_2 = p - uS$.

THEOREM 7. *In the non-singular case, any pair of bilinear forms in $n + n$ variables can be reduced to the classic canonical pair*

$$\alpha = \left(m_1 \sum_{i=1}^{r} \xi_i x_i + v \sum_{i=2}^{r} \xi_i x_{i-1}\right)$$

$$+ \left(m_2 \sum_{i=1}^{s} \eta_i y_i + v \sum_{i=2}^{s} \eta_i y_{i-1}\right) + \cdots,$$

§ 59]   PENCILS OF BILINEAR FORMS   119

$$\beta = \left(n_1 \sum_{i=1}^{r} \xi_i x_i - u \sum_{i=2}^{r} \xi_i x_{i-1}\right)$$
$$+ \left(n_2 \sum_{i=1}^{s} \eta_i y_i - u \sum_{i=2}^{s} \eta_i y_{i-1}\right) + \cdots$$

The elementary divisors of $\rho\alpha + \sigma\beta$ are

$$(m_1 \rho + n_1 \sigma)^r, \quad (m_2 \rho + n_2 \sigma)^s, \ldots$$

COROLLARY. There exists a pair of bilinear forms α and β in $n + n$ variables such that $\rho\alpha + \sigma\beta$ has a determinant which is not zero identically in ρ, σ, and has any preassigned elementary divisors the sum of whose degrees is n.

EXERCISES

1. Two quadratic forms with the matrices A and B have as invariants of index 2 the coefficients of the various powers of λ in the determinant of $A - \lambda B$.

2. By means of Theorem 5, write down the three types of canonical pairs of bilinear forms when $n = 3$.

59. Pencils of bilinear forms. Instead of a pair of bilinear forms μ and ν, with coefficients in F, consider the *pencil* $[\mu, \nu]$ composed of all bilinear forms $\rho\mu + \sigma\nu$, where ρ and σ range over F. This pencil coincides with the pencil $[\alpha, \beta]$, where

(5) $\quad \alpha = p\mu + q\nu, \quad \beta = u\mu + v\nu, \quad pv - qu \neq 0.$

The pencil is called non-singular if the determinant of $\rho\mu + \sigma\nu$ is not zero identically in ρ, σ.

A second such pencil, $[\mu_1, \nu_1]$, is equivalent in F to the pencil $[\mu, \nu]$ if and only if the pair μ_1, ν_1 is equivalent to some pair (5) of the pencil $[\mu, \nu]$.

In view of the Lemma and Theorem 2, we have

THEOREM 8. *Let M, N, M_1, N_1 be the matrices of the bilinear forms μ, ν, μ_1, ν_1 with coefficients in any field F. The non-singular pencils $[\mu, \nu]$ and $[\mu_1, \nu_1]$ are equivalent in F if and only if all the invariant factors (or elementary divisors) of $\rho_1 M_1 + \sigma_1 N_1$ can be derived from those of $\rho M + \sigma N$ by the same non-singular linear transformation expressing ρ and σ linearly in terms of ρ_1 and σ_1 with coefficients in F.*

60. The nth roots of a matrix.
We shall need the

LEMMA. *If $a, b, c, \ldots$ are distinct and not zero, and*

$$\psi(x) = (x - a)^r (x - b)^s (x - c)^t \ldots,$$

there exists a polynomial $\chi(x)$ such that $[\chi(x)]^n - x$ is exactly divisible by $\psi(x)$.

For example, if $n = 2$ and $\psi = (x - a)^2$, then

$$\chi = \tfrac{1}{2} a^{-\frac{1}{2}} (x + a), \qquad \chi^2 - x = \tfrac{1}{4} a^{-1} (x - a)^2.$$

To prove the Lemma, consider the expansion (by the binomial theorem) of $\sqrt[n]{x}$ in a series of ascending powers of $x - a$:

$$\sqrt[n]{x} = \sqrt[n]{a} \sqrt[n]{1 + (x - a)/a}$$
$$= \sqrt[n]{a} \left[1 + \frac{x - a}{na} + \frac{(1 - n)(x - a)^2}{2n^2 a^2} + \cdots \right],$$

and write $F(x)$ for the sum of the first r terms, so that $\sqrt[n]{x} - F(x)$ is termwise divisible by $(x - a)^r$. Similarly, let $G(x)$ be the sum of the first s terms of the expansion of $\sqrt[n]{x}$ in a series of powers of $x - b$, so that $\sqrt[n]{x} - G(x)$ is divisible by $(x - b)^s$; etc.

By decomposition into partial fractions (or by expansion in a series of ascending powers of $x - a$), let

$$\frac{F(x)}{\psi(x)} = \frac{A_0}{(x-a)^r} + \cdots + \frac{A_{-1}}{x-a} + R(x)$$

$$\equiv \frac{A(x)}{(x-a)^r} + \frac{(x-a)^r P(x)}{\psi(x)},$$

where the A_i are constants and $R(x)$ is the quotient of a polynomial $P(x)$ by $\psi(x)/(x-a)^r$. Multiplication by $\psi(x)$ shows that the first r terms in the expansion of $\psi(x)A(x)/(x-a)^r$ in a series of ascending powers of $x - a$ have the sum $F(x)$ and hence are the same as the first r terms in the expansion of $\sqrt[n]{x}$. In other words, the difference between this fraction and $\sqrt[n]{x}$ is divisible by $(x-a)^r$.

Determine $B(x)$ from $G(x)$ just as we determined $A(x)$ from $F(x)$. Then the difference between $\sqrt[n]{x}$ and the second term of

$$\chi(x) = \frac{\psi(x)A(x)}{(x-a)^r} + \frac{\psi(x)B(x)}{(x-b)^s} + \cdots$$

is divisible by $(x-b)^s$. The first, third, ... terms are divisible by $(x-b)^s$ since $\psi(x)$ is. Hence $\chi - \sqrt[n]{x}$ is divisible by $(x-b)^s$ and similarly by $(x-a)^r, \ldots$, and hence by their product $\psi(x)$.

THEOREM 9. *If X is a non-singular square matrix, there exists a polynomial $\chi(\lambda)$ such that the nth power of the matrix $\chi(X)$ is X.*

If $\phi(\lambda)$ is the characteristic determinant of X, then $\phi(X) = 0$ by §25. Since X is non-singular, the constant term of $\phi(\lambda)$ is not zero. Hence we may employ $\phi(\lambda)$ as the $\psi(\lambda)$ of the Lemma and conclude the existence of polynomials $\chi(\lambda)$ and $q(\lambda)$ such that

$$[\chi(\lambda)]^n - \lambda \equiv \phi(\lambda)q(\lambda),$$

identically in λ. Since the coefficients of like powers of λ in each member are equal, and since $X^i X^j = X^{i+j}$, we conclude that

$$[\chi(X)]^n - X = \phi(X)q(X) = 0.$$

61. Equivalence of pairs of quadratic or Hermitian forms, or symmetric or Hermitian bilinear forms. We saw in §56 that the theory of equivalence of pairs of bilinear forms may be developed rationally in any field. But this is not true of pairs of quadratic forms, in the investigation of which we shall employ Theorem 9 and hence introduce irrationalities. Whatever method be employed, irrationalities cannot always be avoided, as is clear from the following example. The real quadratic forms $x^2 - 2y^2$ and $x^2 - 3y^2$ are transformed into $z^2 + 2w^2$ and $z^2 + 3w^2$ respectively by the transformation $x = z$, $y = (-1)^{\frac{1}{2}} w$, but by no real transformation, since the first forms have real linear factors while the second forms do not. We shall therefore study equivalence with respect to the field of all complex numbers.

Consider four bilinear forms α, β, γ, ∂ whose matrices A, B, C, D all have the property $X' = \tilde{X}$. By Theorem 7 of Ch. III, the pair α, β is equivalent to the pair γ, ∂ under a non-singular transformation on the x's with matrix E and the transformation on the y's with matrix $\tilde{E}$ if and only if

(6) $\qquad C = E' A \tilde{E}, \qquad D = E' B \tilde{E}, \qquad |E| \neq 0.$

According as $\tilde{X}$ denotes X or $\bar{X}$, our forms are symmetric or Hermitian bilinear forms. If we identify y_j with $\bar{x}_j$ for $j = 1, \ldots, n$, we conclude that two pairs of quadratic or Hermitian forms are equivalent if and only if their matrices satisfy (6), where $\tilde{E} = E$ or $\bar{E}$ in the respective cases.

THEOREM 10. *If the four n-rowed square matrices A, B, C, D have the property $X' = \tilde{X}$, and if there exist non-singular n-rowed matrices P and Q such that*

(7) $\qquad C = PAQ, \qquad D = PBQ,$

then there exists a non-singular n-rowed matrix E for which equations (6) hold.

By hypothesis,

(8) $\qquad C = \tilde{C}' = \tilde{Q}'\,\tilde{A}'\,\tilde{P}' = \tilde{Q}'\,A\tilde{P}'.$

Equating this to $C = PAQ$, we get

$$(\tilde{Q}')^{-1} PA = A\tilde{P}'\, Q^{-1}.$$

Write

(9) $\qquad X = (\tilde{Q}')^{-1} P,$

so that X is non-singular. Then

$$\tilde{X}' = \tilde{P}'\, Q^{-1},$$

whence

$$XA = A\tilde{X}', \qquad X^2 A = XA\tilde{X}' = A\,(\tilde{X}')^2.$$

By induction on k, we get

(10) $\qquad X^k A = A\,(\tilde{X}')^k.$

Let $\chi(x) = \sum c_k\, x^k$ be any polynomial in x. Multiply $A = A$ by c_0, and (10) by c_k for $k = 1, 2, \ldots$, and add; we get

(11) $\qquad \chi(X)A = A\chi(\tilde{X}').$

By Theorem 9, we may choose complex numbers c_k so that the square of the matrix $V = \chi(X)$ is equal to X. Since X is non-singular and $V^2 = X$, also V is non-singular. Then (11) gives

$$A = V^{-1} A \widetilde{V}'.$$

Thus (7_1) becomes

$$C = P V^{-1} A \widetilde{V}' Q.$$

Write $E = V'\widetilde{Q}$. From $V^2 = X$ and (9), we get $\widetilde{Q}' V = PV^{-1}$. Hence $C = E' A \widetilde{E}$. This proves the first equation (6).

Since X and V and hence also E depend only on P and Q, but not on A or C, the second equation (6) follows similarly from the second equation (7).

Theorems 10 and 3 evidently imply

Theorem 11. *In the non-singular case, two pairs of symmetric (or Hermitian) bilinear forms are equivalent under a non-singular linear transformation on the x's and the cogredient (or conjugate) transformation on the y's if and only if they have the same invariant factors.*

From this and Corollaries 1 and 2 of §§33, 34, we have

Theorem 12. *In the non-singular case, two pairs of quadratic or Hermitian forms are equivalent under non-singular transformation if and only if they have the same invariant factors.*

Consider the case in which the second quadratic or Hermitian form of each pair is $\sum x_i \tilde{x}_i$, whose matrix is the identity matrix I. Hence Theorem 12 implies the

Corollary. Two quadratic (or Hermitian) forms with the matrices A and C are equivalent under an orthogonal (or unitary) transformation if and only if the λ-matrices $A - \lambda I$ and $C - \lambda I$ have the same invariant factors.

The conditions that two pairs of real quadratic forms shall be equivalent under real transformation are complicated.[1]

EXERCISES

1. Every binary non-singular quadratic form is equivalent under an orthogonal transformation (not necessarily real) to one of the following:
$$ax^2 + by^2 (a \neq b); \quad ax^2 + ay^2; \quad 2ax^2 + 2aixy,$$
whose invariant factors are $(a - \lambda)(b - \lambda); \lambda - a, \lambda - a; (a - \lambda)^2$.

2. In the non-singular case, two quadratic forms f and h can be reduced to forms involving only squares of the variables if and only if their elementary divisors are all of the first degree. Hence the latter is true by §37 when f is a real quadratic form and h is positive and non-singular. State the corresponding theorems for Hermitian forms.

62. Pairs involving alternate forms. A bilinear form with the matrix A is called *alternate* if $A' = -A$, and *alternate Hermitian* if $A' = -\bar{A}$ (cf. Exs. 4, 5 of §38).

THEOREM 13. *Theorem* 10 *holds also when*
$$A' = -\tilde{A}, \quad C' = -\tilde{C}, \quad B' = \pm \tilde{B}, \quad D' = \pm \tilde{D}.$$

For, (7_1) then implies
$$C = -\tilde{C}' = -\tilde{Q}' \tilde{A}' \tilde{P}' = +Q' A \tilde{P}',$$
so that (8) holds also here. The rest of the proof of Theorem 10 applies here unchanged.

Theorems 13 and 3 imply

THEOREM 14. *In the non-singular case, two pairs of forms α, β and γ, ∂, of which α and γ are alternate bilinear forms (or alternate Hermitian forms), while β and ∂ are either alternate or are symmetric (or Hermitian) bilinear forms, are equivalent under a non-singular transformation on the x's and the cogredient (or conjugate) transformation on the y's if and only if they have the same invariant factors.*

[1] Muth, Jour. für Math., 128, 1905, 302-21. For pairs of binary quadratic forms in any field, see Dickson, Amer. Jour. Math., 31, 1909, 103-8.

126   PAIRS OF FORMS   [Ch. VI

If α is a symmetric and β an alternate bilinear form, any elementary divisor of $\rho\alpha + \sigma\beta$ of one of the special types ρ^{2k} or σ^{2k+1} is one of an even number of equal elementary divisors.[1]

63. Pairs of forms with preassigned invariant factors.

THEOREM 15. *There exists a pair of quadratic (or Hermitian) forms ϕ and ψ with coefficients in the field F such that ψ is non-singular and the matrix of $\phi - \lambda\psi$ has any preassigned invariant factors I_s with coefficients in F (or in the real subfield S of F), where I_s divides I_{s+1} for every s.*

If there are t invariant factors I_s distinct from unity, we shall prove that we may take $\phi = \sum \alpha_s$, $\psi = \sum \beta_s$, where $\alpha_s - \lambda\beta_s$ has the single invariant factor I_s distinct from unity, and where no variable of α_s or β_s occurs also in α_i or β_i for $i \neq s$ (cf. the proof in §43). In the further discussion we shall omit the subscript s.

For an odd number $2n - 1$ of variables, we may take

$$(12) \begin{cases} \alpha = \sum_{i=1}^{n} a_{ii}\, x_i\, \tilde{x}_i + \sum_{i=1}^{2n-2} x_{i+1}\, \tilde{x}_{2n-i} \\ \quad + \sum_{i=1}^{n-1} (a_{i\ i+1}\, x_i\, \tilde{x}_{i+1} + a_{i+1\ i}\, x_{i+1}\, \tilde{x}_i), \\ \beta = \sum_{i=1}^{2n-1} x_i\, \tilde{x}_{2n-i}, \qquad\qquad a_{ji} = \tilde{a}_{ij}. \end{cases}$$

We shall prove that the determinant Δ_n of $\alpha - \lambda\beta$ has the value

$$\Delta_n = (-1)^{n-1}\Big[-\lambda^{2n-1} + \sum_{i=1}^{n} a_{ii}\, \lambda^{2i-2}$$

$$+ \sum_{i=2}^{n} (a_{i-1\ i} + a_{i\ i-1})\lambda^{2i-3}\Big].$$

---

[1] Kronecker, Monatsber. Akad. Berlin, 1874, 397; Werke, I, 423. Proof by Frobenius in Encyclopédie des Sc. Math., t. I, vol. II, 463–9; and in Muth's Elementartheiler, 1899, 135–142, 231–2. Stickelberger, Jour. für Math., 86, 1879, 42–43.

§ 63] PAIRS WITH ANY INVARIANT FACTORS

We have

$$\Delta_n = \begin{vmatrix} a_{11} & a_{12} & 0 & 0 & \cdots & 0 & 0 & -\lambda \\ a_{21} & a_{22} & a_{23} & 0 & \cdots & 0 & -\lambda & 1 \\ 0 & a_{32} & a_{33} & a_{34} & \cdots & -\lambda & 1 & 0 \\ \multicolumn{8}{c}{\dotfill} \\ 0 & -\lambda & 1 & 0 & \cdots & 0 & 0 & 0 \\ -\lambda & 1 & 0 & 0 & \cdots & 0 & 0 & 0 \end{vmatrix},$$

in which every element of the secondary diagonal is $-\lambda$, except the middle element $a_{nn} - \lambda$, while every element just below that diagonal is 1, and the further elements below it are zero. Hence the minor of a_{11} is ± 1. Multiply the first row by $1/\lambda$ and add to the second row; then multiply the first column by $1/\lambda$ and add to the second column. Now $-\lambda$ is the only element $\neq 0$ in the last row and last column. Hence $\Delta_n = -\lambda^2 d$, where d is the determinant obtained from Δ_n by deleting the first and last rows and columns and replacing a_{22} by

$$b_{22} = a_{22} + (a_{12} + a_{21})/\lambda + a_{11}/\lambda^2.$$

We see that d is derived from Δ_{n-1} by replacing every a_{ij} by $a_{i+1,\,j+1}$ except a_{11}, which is replaced by b_{22}. In a proof of the formula for Δ_n by induction on n, we assume the formula when n is replaced by $n-1$. In the latter we make the preceding replacements and hence get d; its product by $-\lambda^2$ is seen to reduce to the expression for Δ_n. Finally, $\Delta_1 = a_{11} - \lambda$ is the determinant of $(a_{11} - \lambda)x_1\tilde{x}_1$, to which $\alpha - \lambda\beta$ reduces when $n = 1$. Hence the induction is complete.

Hence the matrix $\alpha - \lambda\beta$ has the invariant factors $1, \ldots, 1$, $\pm\Delta_n$. By choice of the a_{ij}, $\pm\Delta_n$ can be identified with any given polynomial in λ of degree $2n - 1$ whose leading coefficient is 1 and the remaining coefficients g have the property $\tilde{g} = g$.

For an even number $2n$ of variables we employ

(13) $\begin{cases} \alpha = \sum_{i=1}^{n} a_{ii} x_i \tilde{x}_i \\ + \sum_{i=1}^{n-1} (a_{i\ i+1} x_i \tilde{x}_{i+1} + a_{i+1\ i} x_{i+1} \tilde{x}_i) \\ + \sum_{i=1}^{2n-1} x_{i+1} \tilde{x}_{2n+1-i} + g x_1 \tilde{x}_{2n} + \tilde{g} x_{2n} \tilde{x}_1, \\ \beta = \sum_{i=1}^{2n} x_i \tilde{x}_{2n+1-i}. \end{cases}$

For $n = 1$, the determinant of $\alpha - \lambda\beta$ is

(14) $\quad \begin{vmatrix} a_{11} & g - \lambda \\ \tilde{g} - \lambda & 1 \end{vmatrix} = -[\lambda^2 - (g + \tilde{g})\lambda + g\tilde{g} - a_{11}].$

For $n > 1$, the determinant Δ of $\alpha - \lambda\beta$ is

$\begin{vmatrix} a_{11} & a_{12} & 0 & \cdots & 0 & 0 & 0 & 0 & 0 & 0 & \cdots & 0 & 0 & g-\lambda \\ a_{21} & a_{22} & a_{23} & \cdots & 0 & 0 & 0 & 0 & 0 & 0 & \cdots & 0 & -\lambda & 1 \\ 0 & a_{32} & a_{33} & \cdots & 0 & 0 & 0 & 0 & 0 & 0 & \cdots & -\lambda & 1 & 0 \\ \multicolumn{14}{c}{\cdots\cdots\cdots\cdots\cdots\cdots\cdots\cdots\cdots\cdots\cdots\cdots\cdots} \\ 0 & 0 & 0 & \cdots & a_{n-1\,n-2} & a_{n-1\,n-1} & a_{n-1\,n} & 0 & -\lambda & 1 & \cdots & 0 & 0 & 0 \\ 0 & 0 & 0 & \cdots & 0 & a_{n\,n-1} & a_{nn} & -\lambda & 1 & 0 & \cdots & 0 & 0 & 0 \\ 0 & 0 & 0 & \cdots & 0 & 0 & -\lambda & 1 & 0 & 0 & \cdots & 0 & 0 & 0 \\ \multicolumn{14}{c}{\cdots\cdots\cdots\cdots\cdots\cdots\cdots\cdots\cdots\cdots\cdots\cdots\cdots} \\ 0 & -\lambda & 1 & \cdots & 0 & 0 & 0 & 0 & 0 & 0 & \cdots & 0 & 0 & 0 \\ \tilde{g}-\lambda & 1 & 0 & \cdots & 0 & 0 & 0 & 0 & 0 & 0 & \cdots & 0 & 0 & 0 \end{vmatrix}$

Multiply the $(n + 1)$th row by λ and add to the nth row; then multiply the $(n + 1)$th column by λ and add to the nth column; we get a like determinant having a_{nn} replaced by $b_{nn} = a_{nn} - \lambda^2$, and having the terms $-\lambda$ in the nth and $(n + 1)$th rows replaced by 0. If $n = 2$, we have (16) with $b_{22} = a_{22} - \lambda^2$.

If $n > 2$, multiply the nth row by λ and add to the $(n-1)$th row; then multiply the nth column by λ and add to the $(n-1)$th column; we get a like determinant having $a_{n-1\ n-1}$ replaced by

(15) $\qquad b_{n-1\ n-1} = a_{n-1\ n-1} + \lambda(a_{n-1\ n} + a_{n\ n-1}) + \lambda^2 b_{nn}$,

having λb_{nn} added to both $a_{n\ n-1}$ and $a_{n-1\ n}$ (which does not effect our final result), and having the terms $-\lambda$ in the $(n-1)$th and $(n+2)$th rows replaced by 0. Thus Δ is the negative of the determinant obtained by deleting the rows and columns numbered n, $n+1$, and $n+2$. However, we shall continue to number the rows as in Δ. If $n = 3$, we have (16), where b_{22} is given by (15).

If $n > 3$, multiply the $(n-1)$th row by λ and add to the $(n-2)$th row; multiply the $(n-1)$th column by λ and add to the $(n-2)$th column; we get a like determinant having $a_{n-2\ n-2}$ replaced by

$$b_{n-2\ n-2} = a_{n-2\ n-2} + \lambda(a_{n-1\ n-2} + a_{n-2\ n-1}) + \lambda^2 b_{n-1\ n-1},$$

having $\lambda b_{n-1\ n-1}$ added to both $a_{n-1\ n-2}$ and $a_{n-2\ n-1}$, and having the terms $-\lambda$ in the $(n-2)$th and $(n+3)$th rows replaced by 0. Thus Δ is equal to the determinant obtained by deleting the rows and columns numbered $n-1$, n, $n+1$, $n+2$, $n+3$.

Proceeding similarly, we see that

(16) $\qquad \Delta = (-1)^{n-2} \cdot \begin{vmatrix} a_{11} & a_{12} & g-\lambda \\ a_{21} & b_{22} & 1 \\ \tilde{g}-\lambda & 1 & 0 \end{vmatrix}.$

Hence

$$\Delta = (-1)^n[-a_{11} + a_{12}\tilde{g} + a_{21}g - (a_{12} + a_{21})\lambda \\ - b_{22}(g-\lambda)(\tilde{g}-\lambda)],$$

where b_{22} may be computed by means of the recursion formula

$$b_{ii} = a_{ii} + \lambda(a_{i\ i+1} + a_{i+1\ i}) + \lambda^2 b_{i+1\ i+1} \quad (i = 2, \ldots, n-1),$$

and $b_{nn} = a_{nn} - \lambda^2$. Thus

$$b_{22} = -\lambda^{2n-2} + \sum_{i=2}^{n} a_{ii} \lambda^{2i-4} + \sum_{i=2}^{n-1} (a_{i\ i+1} + a_{i+1\ i})\lambda^{2i-3}.$$

Hence we obtain the explicit formula

(17) $\quad (-1)^n \Delta = \lambda^{2n} - (g + \tilde{g})\lambda^{2n-1} + (g\tilde{g} - a_{nn})\lambda^{2n-2}$

$\qquad + [-a_{n-1\ n} - a_{n\ n-1} + (g + \tilde{g})a_{nn}]\lambda^{2n-3}$

$\qquad - \sum_{i=2}^{n-1} [a_{ii} - (g + \tilde{g})(a_{i\ i+1} + a_{i+1\ i})$

$\qquad\qquad\qquad + g\tilde{g} a_{i+1\ i+1}]\lambda^{2i-2}$

$\qquad - \sum_{i=3}^{n-1} [a_{i-1\ i} + a_{i\ i-1} + g\tilde{g}(a_{i\ i+1} + a_{i+1\ i}) - a_{ii}(g + \tilde{g})]\lambda^{2i-3}$

$\qquad + [-a_{12} - a_{21} + (g + \tilde{g})a_{22} - g\tilde{g}(a_{23} + a_{32})]\lambda$

$\qquad - a_{11} + a_{12}\tilde{g} + a_{21}g - g\tilde{g}\ a_{22},$

where, by (14), only the first three terms are retained if $n = 1$, while if $n = 2$ we retain the first three and last two terms, and omit the last part of the coefficient of λ.

We can identify (17) with any polynomial $\lambda^{2n} + \cdots$ with coefficients in F or S, according as $\tilde{x}$ denotes x or $\bar{x}$. This may be done by choice in turn of the letters written below the proper powers of λ:

$\lambda^{2n-1},\quad \lambda^{2n-2},\quad\quad \lambda^{2n-3},\ldots,\quad\quad\quad \lambda^{2j-1},\quad\quad \lambda^{2j-2},\ldots$
$-g - \tilde{g},\ -a_{nn},\ -a_{n-1\ n} - a_{n\ n-1},\ldots,\ -a_{j\ j+1} - a_{j+1\ j},\ -a_{jj},\ldots$

where j takes values decreasing to 1 inclusive. We have here omitted from the complete coefficient of any λ^i the terms involving only letters appearing under $\lambda^{2n-1}, \ldots, \lambda^{i+1}$ in the table.

64. Weierstrass's canonical pair of quadratic forms. Whether the number of variables is odd or even, we have exhibited a pair α, β of forms (12) or (13) whose invariant factors are $1, \ldots, 1, I$, where I is any preassigned polynomial. Omitting the signs $\sim$ over x, we have the case of quadratic forms. Let

$$I = \prod_{j=1}^{k} (\lambda - \lambda_j)^{e_j}.$$

If each c_j and each d_j is not zero, the pair

(18)
$$\begin{cases} \gamma = \sum_{j=1}^{k} \left[\lambda_j c_j \sum_{i=1}^{e_j} x_{ij} x_{e_j-i+1,\,j} + d_j \sum_{i=1}^{e_j-1} x_{ij} x_{e_j-i,\,j} \right], \\ \partial = \sum_{j=1}^{k} c_j \sum_{i=1}^{e_j} x_{ij} x_{e_j-i+1,\,j} \end{cases}$$

has the invariant factors $1, \ldots, 1, I$. For, the matrix of $\gamma - \lambda\partial$ has the abbreviated notation

$$\begin{bmatrix} M_1 & 0 & 0 & \cdots & 0 & 0 \\ 0 & M_2 & 0 & \cdots & 0 & 0 \\ \multicolumn{6}{c}{\dotfill} \\ 0 & 0 & 0 & \cdots & 0 & M_k \end{bmatrix},$$

where 0 is a zero matrix, and M_j is the e_j-rowed square matrix of those terms of $\gamma - \lambda\partial$ in which j has a fixed value:

$$M_j = \begin{bmatrix} 0 & 0 & 0 & \cdots & 0 & d_j & c_j(\lambda_j - \lambda) \\ 0 & 0 & 0 & \cdots & d_j & c_j(\lambda_j - \lambda) & 0 \\ \multicolumn{7}{c}{\dotfill} \\ d_j & c_j(\lambda_j - \lambda) & 0 & \cdots & 0 & 0 & 0 \\ c_j(\lambda_j - \lambda) & 0 & 0 & \cdots & 0 & 0 & 0 \end{bmatrix}.$$

The elements of the secondary diagonal are all $c_j(\lambda_j - \lambda)$, those just to the left of them are all d_j, while all further elements are zero. Apart from constant factors, the determinant of M_j is $(\lambda - \lambda_j)^{e_j}$ and the minor of the last element of the last row is $d_j{}^{e_j-1} =$ constant $\neq 0$. As in the first part of §53, the invariant factors of $\gamma - \lambda \partial$ are $1, \ldots, 1, I$. Hence the pair γ, ∂ is equivalent to the pair α, β.

In §63, we obtained a pair of quadratic forms $\phi = \sum \alpha_s$, $\psi = \sum \beta_s$, having exactly t prescribed invariant factors $I_s \neq 1$, where $\alpha_s - \lambda \beta_s$ has the invariant factors $1, \ldots, 1, I_s$. In place of α_s, β_s, we have now found that we may take a pair γ_s, ∂_s of type (18). Hence $\sum \gamma_s, \sum \partial_s$ have the t prescribed invariant factors $\neq 1$. Changing the meaning of k, we may now suppose that $(\lambda - \lambda_j)^{e_j}$, for $j = 1, \ldots, k$, give all the highest powers of linear functions which occur among all the invariant factors (instead of those for a single one I as before). Hence we obtain

Theorem 16. *There exists a pair* (18) *of quadratic forms having any prescribed elementary divisors*

$$(\lambda - \lambda_1)^{e_1}, \ldots, (\lambda - \lambda_k)^{e_k}.$$

For example, let the number of variables be three.

(i) $e_1 = e_2 = e_3 = 1$. Take $c_1 = c_2 = 1$, $c_3 = -1$, and write x_j for x_{1j}. Thus

$$\gamma = \lambda_1 x_1{}^2 + \lambda_2 x_2{}^2 - \lambda_3 x_3{}^2, \qquad \partial = x_1{}^2 + x_2{}^2 - x_3{}^2.$$

If the λ's are distinct, the conics $\gamma = 0$ and $\partial = 0$ have four distinct points of intersection given by

$$x_1{}^2 : x_2{}^2 : x_3{}^2 = \lambda_2 - \lambda_3 : \lambda_3 - \lambda_1 : \lambda_2 - \lambda_1.$$

If $\lambda_1 \neq \lambda_2 = \lambda_3$, the only intersections have $x_1 = 0$, $x_2 = \pm x_3$, and the conics are tangent at two points. If $\lambda_1 = \lambda_2 = \lambda_3 \neq 0$, the conics coincide.

(ii) $e_1 = 2$, $e_2 = 1$. Take $c_1 = c_2 = d_1 = 1$, and write x_1, x_2, x_3 for x_{11}, x_{21}, x_{12}. Then

§ 65] LIST OF FURTHER RESULTS 133

$$\gamma = 2\lambda_1 x_1 x_2 + x_1{}^2 + \lambda_2 x_3{}^2, \qquad \partial = 2x_1 x_2 + x_3{}^2.$$

If $\lambda_1 = \lambda_2$, the only intersection has $x_1 = x_3 = 0$, and the conics have contact of the third order. But if $\lambda_1 \neq \lambda_2$, there are two further points of intersection having

$$x_1 = 2(\lambda_2 - \lambda_1)x_2, \qquad x_3{}^2 = -4(\lambda_2 - \lambda_1)x_2{}^2,$$

and the conics have simple contact at $x_1 = x_3 = 0$.

(iii) $e_1 = 3$. Take $c_1 = d_1 = 1$, and omit the second subscripts on the x's. Then

$$\gamma = \lambda_1(2x_1 x_3 + x_2{}^2) + 2x_1 x_2, \qquad \partial = 2x_1 x_3 + x_2{}^2.$$

The conics have contact of the second order at $x_1 = x_2 = 0$, and cross at $x_2 = x_3 = 0$.

EXERCISES

1. List the canonical forms of pairs of quadratic forms in four variables.[1]

2. If A and N are any two n-rowed square matrices, there exists an n-rowed non-singular matrix X such that $X'A\tilde{X} = N$ (whence A and N are congruent or conjunctive) if and only if the pair $A, \tilde{A}'$ is equivalent[2] to the pair $N, \tilde{N}'$. Hints: If they are equivalent, there exist matrices P and Q such that $PAQ = N, P\tilde{A}'Q = \tilde{N}'$. Add and subtract and apply Theorem 13.

65. Further applications of matrices to forms, available in English. Hurwitz's identity expressing the product of a sum of p squares and a sum of n squares as a sum of n squares.[3]

---

[1] Bromwich, Quadratic Forms, 1906, 46; or Hilton, Homogeneous Linear Substitutions, 1914, 105.

[2] Hence the problem depends upon the singular case. Muth, Elementartheiler, 1899, 144–51. Encyclopédie des sc. math., t. I, v. II, 471–4.

[3] For $p = n$, Dickson, Annals of Math., (2), 20, 1919, 160–4. Hurwitz, Math. Annalen, 88, 1922, 1–25. Radon, Abhand. Math. Hamburg Univ., 1, 1921, 1–14, found the maximum p for a given n when the matrices are real. Generalization to forms of any order, Dickson, Congrès International des Mathématiciens, Strasbourg, 1920, 215–30; Comptes Rendus Paris, 172, 1921, 636–40, 1262.

Forms expressible as determinants with linear elements.[1]

Frobenius's theory of matrices whose elements are polynomials in one variable; application to the equivalence of pairs of bilinear forms.[2]

Equivalence of two quadratic or Hermitian forms in a general field.[3]

Singular case of pairs of bilinear, quadratic, or Hermitian forms.[4]

Weierstrass[5] gave a simple application of the theory of pairs of bilinear forms to the integration of a system of n linear differential equations of the first order in n variables with constant coefficients.

For applications of elementary divisors to linear differential equations, see texts on the latter subject.

[1] Dickson, Trans. Amer. Math. Soc., 22, 1921, 167–79 (24, 1922, 185; 26, 1926, 367); Amer. Jour. Math., 43, 1921, 102–34; Annals of Math., 23, 1921, 70–74.

[2] Bôcher, Introduction to Higher Algebra, 1907, 262–284. Cf. Dickson, Algebras and their Arithmetics, 1923, 169–174; revised in the German edition, Zürich, 1926.

[3] Dickson, Bull. Amer. Math. Soc., 14, 1907–8, 108–115 (simplification of Trans. Amer. Math. Soc., 7, 1906, 275–292); Quar. Jour. Math., 39, 1908, 316–33.

[4] Kronecker, Berlin. Sitzungsberichte, 1890, 1225–37, 1375–88; 1891, 9–17, 33–44. Simpler treatment by rational methods, Dickson, Trans. Amer. Math. Soc., 1927 (10, 1909, 358–60).

[5] Berlin. Abhandl., 1875; Werke, II, 75–76.

Chapter VII

FIRST PRINCIPLES OF GROUPS OF SUBSTITUTIONS

The fundamental concepts of substitutions and groups will be introduced in a very concrete and natural way in connection with the solution of cubic and quartic equations. The reader will therefore appreciate from the start some of the reasons why these concepts are employed. The deliberate presentation and numerous illustrative examples will enable the reader to digest these somewhat abstract ideas.

66. Cubic equations. Of various methods of solving

(1) $$x^3 + bx^2 + cx + d = 0$$

we shall present the method which best illustrates our later theory of the solution of equations of any degree. If x_1, x_2, x_3 are the roots, then

(2) $$\sum x_1 \equiv x_1 + x_2 + x_3 = -b,$$
$$\sum x_1 x_2 \equiv x_1 x_2 + x_1 x_3 + x_2 x_3 = c, \qquad x_1 x_2 x_3 = -d.$$

Let ω be an imaginary cube root of unity, so that

(3) $$\omega^2 + \omega + 1 = 0, \qquad \omega^3 = 1.$$

As soon as we have computed the values of

(4) $$\phi = x_1 + \omega x_2 + \omega^2 x_3, \qquad \psi = x_1 + \omega^2 x_2 + \omega x_3,$$

we can find the x's by solving these linear equations with the first one of (2). First, we add the three equations and get the value of $3x_1$. Next, we multiply them by ω^2, ω, 1, respectively, and add. Finally, we multiply them by ω, ω^2, 1 and add. Using (3), we get

(5) $$x_1 = \tfrac{1}{3}(\phi + \psi - b), \quad x_2 = \tfrac{1}{3}(\omega^2 \phi + \omega\psi - b),$$
$$x_3 = \tfrac{1}{3}(\omega\phi + \omega^2 \psi - b).$$

To compute the values of ϕ and ψ, we employ

(6) $$\phi\psi = \sum x_1^2 + (\omega + \omega^2)\sum x_1 x_2$$
$$= (\sum x_1)^2 - 3\sum x_1 x_2 = A = b^2 - 3c,$$

(7) $$\phi^3 + \psi^3 = 2\sum x_1^3 - 3\sum x_1^2 x_2 + 12 x_1 x_2 x_3$$
$$= 2(\sum x_1)^3 - 9\sum x_1 \cdot \sum x_1 x_2 + 27 x_1 x_2 x_3 = B,$$
$$B = -2b^3 + 9bc - 27d,$$
$$(\phi^3 - \psi^3)^2 = (\phi^3 + \psi^3)^2 - 4\phi^3 \psi^3 = B^2 - 4A^3.$$

Hence $\phi^3 - \psi^3$ is equal to one of the square roots of $B^2 - 4A^3$. Choosing a particular square root and employing also (7), we get

(8) $$\phi^3 = \tfrac{1}{2}[B + (B^2 - 4A^3)^{\frac{1}{2}}] = g,$$
$$\psi^3 = \tfrac{1}{2}[B - (B^2 - 4A^3)^{\frac{1}{2}}] = h.$$

The cube roots ϕ and ψ of g and h must be chosen so that $\phi\psi = A$ by (6). We agree to obtain ϕ as a cube root of g and then compute ψ as the quotient A/ϕ.

If ϕ is a particular cube root of g, the remaining cube roots are $\omega\phi$ and $\omega^2\phi$. We therefore consider the functions

(9) $$\phi_1 = \omega\phi = x_3 + \omega x_1 + \omega^2 x_2,$$
$$\psi_1 = \omega^2 \psi = x_3 + \omega x_2 + \omega^2 x_1,$$

(10) $$\phi_2 = \omega^2 \phi = x_2 + \omega x_3 + \omega^2 x_1,$$
$$\psi_2 = \omega\psi = x_2 + \omega x_1 + \omega^2 x_3.$$

The replacement of ϕ and ψ in (5) by $\omega\phi$ and $\omega^2\psi$, respectively, has the effect of replacing x_1 by x_3, x_2 by x_1, and x_3 by x_2. This operation is called a *substitution* on the roots (or letters) x_1, x_2, x_3, and is denoted by $(x_1 \, x_3 \, x_2)$. Such a *cycle* signifies that each letter is replaced by the letter written just to the right of it, while the last letter x_2 of the cycle is replaced by the first letter x_1.

Similarly, the replacement of ϕ and ψ in (5) by $\omega^2\phi$ and $\omega\psi$ respectively has the effect of applying the substitution $(x_1 \, x_2 \, x_3)$.

If we had chosen the other square root of $B^2 - 4A^3$, we would have obtained $\phi^3 = h$, $\psi^3 = g$, instead of (8). The interchange of ϕ and ψ in (5) has the effect of applying the substitution $(x_2 \, x_3)$ which interchanges x_2 with x_3 and leaves x_1 unaltered. The replacement of ϕ by $\omega\psi$ and ψ by $\omega^2\phi$ in (5) has the effect of applying the substitution $(x_1 \, x_2)$. Finally, the replacement of ϕ by $\omega^2\psi$ and ψ by $\omega\phi$ in (5) has the effect of applying the substitution $(x_1 \, x_3)$.

We may combine the essential parts of these results as follows: If we have found a definite value of ϕ (and hence of $\psi = A/\phi$) by making a choice of the square root in (8) and then a choice of the cube root of the resulting g, so that formulas (5) give definite values for x_1, x_2, x_3, then a different choice of the square root or of the cube root leads to new values of x_1, x_2, x_3 which are derived from the former values by applying a substitution on the roots. This conclusion might have been anticipated, since if we start with the coefficients of the cubic equation and compute the roots, we are at liberty to assign the notations x_1, x_2, x_3 to the roots arranged in any order.

Exercise

Employing the linear functions (4), (9), (10), verify that the substitutions $(x_1 \, x_3 \, x_2)$ and $(x_1 \, x_2 \, x_3)$ replace ϕ and ψ by ϕ_1 and ψ_1, and ϕ_2 and ψ_2, respectively, while $(x_2 \, x_3)$ interchanges ϕ and ψ, $(x_1 \, x_2)$ interchanges ϕ_1 and ψ_1, and $(x_1 \, x_3)$ interchanges ϕ_2 and ψ_2. Also, $(x_1 \, x_2)$ replaces ϕ and ψ by ψ_2 and ϕ_2, while $(x_1 \, x_3)$ replaces ϕ and ψ by ψ_1 and ϕ_1, respectively.

67. Discriminant of a cubic equation. The discriminant Δ of a cubic equation in which the coefficient of x^3 is unity is defined to be the product of the squares of the differences of the roots. We shall compute it by means of the value $B^2 - 4A^3$ of $(\phi^3 - \psi^3)^2$ in §66. Since the cube roots of unity are 1, ω, and ω^2, we have

$$z^3 - 1 \equiv (z - 1)(z - \omega)(z - \omega^2)$$

for all values of z. Taking $z = \phi/\psi$, we see that

$$\phi^3 - \psi^3 = (\phi - \psi)(\phi - \omega\psi)(\phi - \omega^2\psi).$$

From (4), (9), and (10), we get

(11)
$$\phi - \psi = (\omega - \omega^2)(x_2 - x_3),$$
$$\phi - \omega\psi = (1 - \omega)(x_1 - x_2),$$
$$\phi - \omega^2\psi = (1 - \omega^2)(x_1 - x_3).$$

We desire the product of the squares of the three expressions (11). By use of (3), we get

$$(\omega - \omega^2)^2 = -3, \qquad (1 - \omega)(1 - \omega^2) = 3.$$

Hence $B^2 - 4A^3 = -27\Delta$. This gives

(12) $\qquad \Delta = 18bcd - 4b^3 d + b^2 c^2 - 4c^3 - 27d^2.$

EXERCISES

1. The discriminant of $x^3 + cx + d = 0$ is $-4c^3 - 27d^2$.

2. Show that $x^3 - 27x + 54 = 0$ has a multiple root by computing its discriminant.

3. If we define the discriminant Δ of $ax^3 + bx^2 + cx + d = 0$ to be $a^4 P$, where P is the product of the squares of the differences of the roots, so that P is derived from (12) by replacing b, c, d by $b/a, c/a, d/a$, respectively, show that $\Delta = 18abcd - 4b^3 d + b^2 c^2 - 4ac^3 - 27a^2 d^2$.

4. If D is the discriminant of the corresponding cubic form $ax^3 + bx^2 y + cxy^2 + dy^3$, show by (19) of Ch. I that $\Delta = -27D$.

5. If $\Delta \neq 0$ and $A \neq 0$ (so that $\phi \neq 0$, $\psi \neq 0$), verify by using (11) that the six functions (4), (9), and (10) have six distinct values. In the Exercise in §66, we gave a substitution which replaces ϕ by any one of the remaining five functions. The *identity* substitution which replaces each root x_i by itself does not of course alter ϕ. Since therefore the six substitutions on the three roots replace ϕ by six functions whose values are all distinct, ϕ is called a six-valued function.

6. Show that also ψ is a six-valued function. Why does it follow without computation that each of the four functions (9) and (10) is six-valued?

68. Substitutions on n letters. The operation s which replaces x_1 by x_a, x_2 by x_b, ..., x_n by x_l, where $a, b, \ldots, l$ form a permutation of $1, 2, \ldots, n$, is called a *substitution* on $x_1, x_2, \ldots, x_n$. It is given the two-rowed notation

$$(13) \qquad s = \begin{pmatrix} x_1 \; x_2 \; \cdots \; x_n \\ x_a \; x_b \; \cdots \; x_l \end{pmatrix} = \begin{pmatrix} 1 \; 2 \; \cdots \; n \\ a \; b \; \cdots \; l \end{pmatrix},$$

where in the second form only the subscripts are written. Since the order of the columns is immaterial, we may permute them in any manner without changing the meaning of s.

To each permutation $a, b, \ldots, l$ of $1, 2, \ldots, n$ corresponds one and only one substitution. Hence *the number of distinct substitutions on n letters is $n! = n(n-1) \cdots 3 \cdot 2 \cdot 1$.*

Instead of the two-rowed notation in (13), we often employ the one-rowed notation of cycles (§66). The relation between the two notations will be clear from the following substitutions:

$$\begin{pmatrix} 1 \; 2 \; 3 \; 4 \; 5 \\ 2 \; 3 \; 1 \; 4 \; 5 \end{pmatrix} = (123), \qquad \begin{pmatrix} 1 \; 2 \; 3 \; 4 \; 5 \\ 2 \; 3 \; 1 \; 5 \; 4 \end{pmatrix} = (123)(45).$$

69. Product of substitutions. Given a substitution s in (13) and another substitution t, we may permute the columns of t in such a way that the letters in its upper row are the same as the corresponding letters in the lower row of s. Then if

(14) $$t = \begin{pmatrix} a & b & \cdots & l \\ \alpha & \beta & \cdots & \lambda \end{pmatrix},$$

the effect of applying first s and afterwards t is the same as the effect of applying the single substitution

(15) $$p = \begin{pmatrix} 1 & 2 & \cdots & n \\ \alpha & \beta & \cdots & \lambda \end{pmatrix},$$

which is called the *product* of s by t. We write $p = st$. The word product is used here in the sense of resultant or compound.

We may, of course, find the product of substitutions each expressed in cycles. For example,

$$(12)(34) \cdot (23) = (1342), \quad (23) \cdot (12)(34) = (1243).$$

To find the first product, note that the first factor replaces 1 by 2 and the second factor replaces 2 by 3, whence the product replaces 1 by 3; etc. Since the two products of our substitutions taken in different orders are distinct, we see that multiplication of substitutions is not always commutative.

When a substitution is composed of two or more cycles (on different letters), it may be regarded as a product of commutative cyclic substitutions each composed of a single cycle. For example, $(12)(34) = (12) \cdot (34) = (34) \cdot (12)$.

Multiplication of substitutions is *associative*: $st \cdot v = s \cdot tv$, so that the notation stv is unambiguous. Let s, t, and the product $st = p$ have the notations in (13), (14), (15). Let

$$v = \begin{pmatrix} \alpha & \beta & \cdots & \lambda \\ A & B & \cdots & L \end{pmatrix}.$$

§ 69] PRODUCT OF SUBSTITUTIONS 141

Then

$$tv = \begin{pmatrix} ab & \cdots & l \\ AB & \cdots & L \end{pmatrix}, \quad st \cdot v = pv = \begin{pmatrix} 1 & 2 & \cdots & n \\ A & B & \cdots & L \end{pmatrix} = s \cdot tv.$$

From this result we readily deduce the associative law for products of four substitutions a, b, c, d:

$$a \cdot bcd = a[b \cdot cd] = ab \cdot cd = [ab \cdot c]d = abc \cdot d,$$

so that the notation $abcd$ is unambiguous. The associative law for products of any number k of substitutions is readily proved similarly by induction on k. Thus there is a single product of k substitutions $a_1, a_2, \ldots, a_k$ taken in a fixed order, and the notation $a_1 a_2 \cdots a_k$ is unambiguous. In particular, if the k factors are all equal to a, their product is denoted by a^k without ambiguity and is called the kth *power* of a.

70. Identity, order, inverse. The substitution which leaves unaltered each of the n letters is called the *identity* and denoted by I.

A substitution p is said to be of *order* (or period) k if k is the least positive integer such that $p^k = I$. For example, (12) and (12)(34) are of order 2.

If $p = (123 \ldots r)$, p^2 replaces 1 by 3, p^3 replaces 1 by 4, ..., and for $j < r$, p^j replaces 1 by $j + 1$, so that the order of p is not less than r. Moreover, $p^r = I$. Hence the order of a single cycle on r letters is r.

If $a, b, \ldots$ are single cycles of $r, s, \ldots$ letters respectively, and if no two of these cycles have a letter in common, then $[ab \cdots]^k = a^k b^k \cdots$ will be the identity if and only if $a^k = I, b^k = I, \ldots$, and hence if and only if k is a common multiple of $r, s, \ldots$. Hence the order of any substitution is the least common multiple of the orders (numbers of letters) of its component cycles. For example, (123)(45) is of order 6, while (12)(3456) is of order 4.

Every substitution s has an *inverse* s^{-1} such that $ss^{-1} = I$, $s^{-1}s = I$. This follows from

(16) $\qquad s = \begin{pmatrix} 1\ 2\ \cdots\ n \\ a\ b\ \cdots\ l \end{pmatrix}, \qquad s^{-1} = \begin{pmatrix} a\ b\ \cdots\ l \\ 1\ 2\ \cdots\ n \end{pmatrix}.$

Exercises

1. If $s = (123)$, verify that $s^2 = (132)$, $s^3 = I$, $s^{-1} = s^2$.
2. If s is of order k, $s^{-1} = s^{k-1}$.
3. A cycle is not altered by a cyclic permutation of its letters.
4. If $st = sr$, then $t = r$.
5. The inverse of $(123 \cdots r)$ is $(r\ \ r-1\ \cdots\ 321)$.

71. Quartic equations. To solve

(17) $\qquad x^4 + bx^3 + cx^2 + dx + e = 0,$

transpose the last three terms and complete the square on the first two; we get

$$(x^2 + \tfrac{1}{2}bx)^2 = (\tfrac{1}{4}b^2 - c)x^2 - dx - e.$$

The device of Ferrari (1522–1565) consists in deriving an equation in which also the second member will be a perfect square. This is accomplished by adding $(x^2 + \tfrac{1}{2}bx)y + \tfrac{1}{4}y^2$ to each member; we get

(18) $\qquad (x^2 + \tfrac{1}{2}bx + \tfrac{1}{2}y)^2 = (\tfrac{1}{4}b^2 - c + y)x^2 + (\tfrac{1}{2}by - d)x + \tfrac{1}{4}y^2 - e.$

Denote the second member by $q \equiv rx^2 + sx + t$. If it is the square of $mx + n$, then $m^2 = r$, $2mn = s$, $n^2 = t$, whence $s^2 = 4rt$. Conversely, let the last condition be satisfied. If $r = 0$,

§ 71] QUARTIC EQUATIONS 143

then $s = 0$ and q is a constant and hence a square. If $r \neq 0$, q is the square of $r^{\frac{1}{2}} x + \frac{1}{2} s r^{-\frac{1}{2}}$. Hence $rx^2 + sx + t$ is the square of a linear function of x if and only if $s^2 = 4rt$. Applying this result to the second member of (18), we see that it is a perfect square if and only if

(19) $\quad y^3 - cy^2 + (bd - 4e)y + 4ce - b^2 e - d^2 = 0.$

If y_1 is a root of this *resolvent cubic equation*, (18) implies

(20) $\quad \begin{aligned} & x^2 + \tfrac{1}{2} bx + \tfrac{1}{2} y_1 = mx + n \\ & \text{or } x^2 + \tfrac{1}{2} bx + \tfrac{1}{2} y_1 = - mx - n. \end{aligned}$

Let x_1 and x_2 be the roots of the first of these quadratic equations, and x_3 and x_4 be the roots of the second. Then $x_1, \ldots, x_4$ are the desired roots of (17). Note that

$$x_1 x_2 = \tfrac{1}{2} y_1 - n, \quad x_3 x_4 = \tfrac{1}{2} y_1 + n, \quad x_1 x_2 + x_3 x_4 = y_1.$$

If instead of y_1 we employ another root y_2 or y_3 of the resolvent cubic (19), we obtain two quadratic equations different from (20), such that their four roots are x_1, x_2, x_3, x_4 paired in a new manner. The root which is paired with x_1 is now x_3 or x_4. Hence the roots of the resolvent cubic (19) are

(21) $\quad y_1 = x_1 x_2 + x_3 x_4, \quad y_2 = x_1 x_3 + x_2 x_4, \quad y_3 = x_1 x_4 + x_2 x_3.$

The discriminant of (17) is defined to be the product of the squares of the differences of its roots. By (21),

(22) $\quad \begin{aligned} y_1 - y_2 &= (x_1 - x_4)(x_2 - x_3), \\ y_1 - y_3 &= (x_1 - x_3)(x_2 - x_4), \\ y_2 - y_3 &= (x_1 - x_2)(x_3 - x_4). \end{aligned}$

144   GROUPS OF SUBSTITUTIONS   [Ch. VII

Taking the product of the squares, we see that *the discriminant of a quartic equation* (17) *is equal to the discriminant of its resolvent cubic* (19).

EXERCISES

1. The resolvent cubic of $x^4 + 1 = 0$ is $y^3 - 4y = 0$. Hence $x^4 + 1$ has the pairs of factors $x^2 \pm i$, $x^2 \pm 2^{\frac{1}{2}} x + 1$, $x^2 \pm (-2)^{\frac{1}{2}} x - 1$.

2. The discriminant of $x^4 - 8x^3 + 22x^2 - 24x + 9 = 0$ is zero, so that the equation has a multiple root.

3. If $d = b, e = 1$, so that (17) is a reciprocal equation, the resolvent cubic is $(y - 2)[y^2 + (2 - c)y + b^2 - 2c] = 0$. The fact that 2 is a value of one of the y's in (21) follows directly from the fact that the roots are reciprocal in pairs. In particular, taking $y = 2$, we find that $x^4 + x^3 + x^2 + x + 1$ has the factors $x^2 + \frac{1}{2}(1 \pm 5^{\frac{1}{2}})x + 1$.

72. Groups of substitutions. A set of m distinct substitutions is called a *group* of *order m* if every product of two of them and the square of each of them are substitutions of the set.

For example, the $n!$ substitutions on n letters form a group, called the *symmetric group* on n letters, since the product of any two substitutions on n letters and the square of any one of them are substitutions on n letters. Again, if s is a substitution of order m, the substitutions $I, s, s^2, \ldots, s^{m-1}$ form a *cyclic group* of order m, *generated by s*. The *identity group* is composed of the single substitution I.

If all the substitutions of a group belong to another group, the former is called a *subgroup* of the latter.

As a generalization, we shall next define a group whose elements need not be substitutions. An *abstract group* is a system composed of a set of elements $a, b, \ldots$ and a rule of combining any two of them to produce their "product," such that (i) every product of two of the elements and the square of each element are elements of the set, (ii) the associative law holds, (iii) the set contains an *identity* element I such that $Ia = aI = a$ for every element a of the set, and (iv) each element a of the set has an inverse a^{-1} belonging to the set, such that $aa^{-1} = a^{-1}a = I$.

In our definition of a group of substitutions, we employed assumption (i) alone. The associative law holds (§69). If a is a substitution of order m of the set, then $a^m = I$ and $a^{m-1} = a^{-1}$ belong to the set. Hence for sets of substitutions, properties (iii) and (iv) follow from (i) and (ii).

73. Group leaving a function invariant. Let $x_1, \ldots, x_4$ be distinct. By (22), the three functions $y_1 = x_1 x_2 + x_3 x_4$, $y_2 = x_1 x_3 + x_2 x_4$, $y_3 = x_1 x_4 + x_2 x_3$ are distinct. Since y_2 and y_3 are the only other functions of the same form as y_1, it is to be anticipated that each of the 24 substitutions on x_1, x_2, x_3, x_4 replaces y_1 by y_1, y_2, or y_3 and hence that $\frac{1}{3}24$ or 8 of them leave y_1 unaltered. To verify these statements, note that $a = (12)$ and $b = (13)(24)$ leave $x_1 x_2 + x_3 x_4$ unaltered. The same is therefore true of the products

$$c = ab = (1423), \quad d = ba = (1324), \quad e = ca = (14)(23),$$
$$be = (12)(34), \quad a \cdot be = (34).$$

Hence y_1 is unaltered by the eight substitutions of the set

$$G_8 = \{I, (12), (34), (12)(34),$$
$$(13)(24), (14)(23), (1423), (1324)\}.$$

Evidently substitution (23) replaces y_1 by y_2. Hence the eight distinct products of the eight substitutions of G_8 by (23) all replace y_1 by y_2. Similarly, the eight distinct products of the eight substitutions of G_8 by (24) all replace y_1 by y_3. There is no substitution in common with two of these three sets of eight substitutions, since they replace y_1 by y_1, y_2, y_3, respectively. Hence together they give all of the 24 substitutions on x_1, x_2, x_3, x_4. We conclude that every substitution on the x's replaces y_1 by y_1, y_2, or y_3, so that y_1 is a three-valued function. Moreover, we conclude that G_8 is composed of all the substitutions on the x's which leave y_1 unaltered. Hence every product of two equal or distinct substitutions of G_8 is equal to a substitution of G_8, so that G_8 is a group of order 8.

We shall say that the function $x_1 x_2 + x_3 x_4$ *belongs* to the group G_8 since it remains unaltered by all the substitutions of G_8 and is altered by all further substitutions on x_1, x_2, x_3, x_4. We saw that all eight substitutions of G_8 may be obtained by successive multiplications starting with a and b. Hence the group G_8 is said to be *generated* by a and b.

Exercises

1. Since $y_2 = x_1 x_3 + x_2 x_4$ is derived from y_1 by interchanging x_2 and x_3, the group H_8 to which y_2 belongs is derived from G_8 by interchanging 2 and 3:

$$H_8 = \{I,\ (13),\ (24),\ (13)(24),\ (12)(34),\ (14)(23),\ (1432),\ (1234)\}.$$

2. By interchanging 2 and 4 in G_8, obtain the group

$$K_8 = \{I,\ (14),\ (23),\ (14)(23),\ (13)(24),\ (12)(34),\ (1243),\ (1342)\}$$

to which $y_3 = x_1 x_4 + x_2 x_3$ belongs. Find two generators of this group.

3. For $t = x_1 + x_2 - x_3 - x_4$, we have $t^2 = 4y_1 + b^2 - 4c$. Hence a substitution leaving t unaltered leaves also y_1 unaltered. Show that t belongs to the subgroup $\{I,\ (12),\ (34),\ (12)(34)\}$ of G_8.

4. The six functions obtained from $\phi = y_1 + \omega y_2 + \omega^2 y_3$ by applying the six substitutions on y_1, y_2, y_3 are all distinct if [1]$x_1, \ldots, x_4$ are independent variables. Then ϕ belongs to the group G composed of the substitutions on the x's which leave y_1, y_2, y_3 simultaneously unaltered. Hence G is the greatest common subgroup of G_8, H_8, K_8, so that

$$G = \{I,\ (12)(34),\ (13)(24),\ (14)(23)\}.$$

Verify also that G is a commutative group.

5. For three distinct x's, $(x_1 - x_2)(x_1 - x_3)(x_2 - x_3)$ belongs to the group $\{I, a = (123), (132)\}$. Verify also that this is a cyclic group generated by a.

74. Alternating group.

A *transposition* is a substitution, like (13), which interchanges two letters and leaves unaltered the further letters. Every substitution is a product of transpositions.

---

[1] And in further cases. See Ex. 5 of §67.

§ 74]   ALTERNATING GROUP   147

For, it is a product of cycles on different letters, while a single cycle with n letters is a product of $n - 1$ transpositions:

$$(123 \cdots n - 1\, n) = (12)(13) \cdots (1\, n - 1)(1\, n).$$

But the same substitution can be decomposed into transpositions in various ways. For example,

$$(123) = (12)(13) = (13)(23) = (13)(23) \cdot (45)(45).$$

Of the various decompositions of a given substitution s into transpositions, either all contain an even number of transpositions (and s is then called an *even* or *positive* substitution), or all contain an odd number of transpositions (and s is called an *odd* or *negative* substitution). This is proved by using the *alternating function*

$$\begin{aligned}
P = (x_1 - x_2)(x_1 - x_3)(x_1 - x_4) &\cdots (x_1 - x_n) \\
\cdot (x_2 - x_3)(x_2 - x_4) &\cdots (x_2 - x_n) \\
\cdot \cdot \cdot \cdot \cdot \cdot \cdot \cdot &\\
&\cdot (x_{n-1} - x_n),
\end{aligned}$$

in which $x_1, \ldots, x_n$ are assumed to be distinct. Any transposition $(x_i\, x_j)$ with $i < j$ merely changes the sign of P (see the preceding Ex. 5). For, the factors of P which involve neither x_i nor x_j are evidently unaltered. The factor $x_i - x_j$ involving both is changed in sign. The remaining factors may be paired to form the products

$$\pm (x_i - x_k)(x_j - x_k) \qquad [k = 1, \ldots, n; k \neq i, k \neq j].$$

Such a product is unaltered. Hence P is merely changed in sign.

Hence if s is a product of an even number of transpositions, it leaves P unaltered. But if s is a product of an odd number of transpositions, it replaces P by $-P$. This proves the above assertion which justifies the terms even and odd substitutions.

All even substitutions on n letters therefore form a group, called the *alternating* group on n letters. It is the group to which the alternating function P belongs.

75. Theorem. *If a group G contains an odd substitution t, exactly half of the substitutions of G are odd.*

For, the products of its even substitutions by t are distinct and odd, so that G contains at least as many odd substitutions as even ones. Again, the products of its odd substitutions by t are distinct and even, so that G contains at least as many even substitutions as odd ones. The two results show that G contains exactly as many odd as even substitutions.

It follows that the alternating group on n letters is of order $\frac{1}{2} \cdot n!$. Since it is a subgroup of the symmetric group of order $n!$, we have an illustration of the next theorem.

76. Theorem. *The order of any group is a multiple of the order of any subgroup.*

Let a group G of order r have a subgroup H of order $s < r$, so that G contains a substitution g_2 not in H. Write Hg_2 for the set of s substitutions obtained by multiplying each substitution of H by g_2. If a substitution hg_2 of Hg_2 were equal to a substitution h' of H, then $g_2 = h^{-1} h'$ would belong to the group H, contrary to hypothesis. Hence there are $2s$ distinct substitutions in the set $H + Hg_2$, which denotes the aggregate of the sets H and Hg_2. If $r = 2s$, the theorem is proved. If $r > 2s$, G contains a substitution g_3 not in H or Hg_2. As before, no substitution of Hg_3 belongs to H. If a substitution hg_3 of Hg_3 were equal to a substitution $h' g_2$ of Hg_2, then $g_3 = h^{-1} h' g_2 = h_1 g_2$ would belong to Hg_2, contrary to hypothesis. Hence there are $3s$ distinct substitutions in the set $H + Hg_2 + Hg_3$. Unless the latter coincides with G, we repeat the argument using a new substitution g_4.

If G is a group on n letters, its order does not exceed $n!$. Hence the above process terminates and gives

$$G = H + Hg_2 + Hg_3 + \cdots + Hg_u.$$

Thus G is of order $r = su$. This integer u is called the *index* of H under G.

EXERCISES

1. The order of any group on n letters is a divisor of $n!$.
2. The order of any substitution of a group G divides the order of G.
3. Hence a group of prime order is cyclic.
4. The alternating group on 4 letters is composed of I, (12)(34), (13)(24), (14)(23), and the eight cycles of three letters.
5. All the substitutions common to two groups form a subgroup of each.

Chapter VIII

FIELDS, REDUCIBLE AND IRREDUCIBLE FUNCTIONS

This chapter presents certain concepts and elementary theorems which will be needed in the Galois theory of equations.

77. Fields, adjunction. A set of complex numbers is called a *number field* if the sum, difference, product, and quotient (except by zero) of any two equal or distinct numbers of the set are themselves numbers of the set. Examples were given in §28.

All rational functions of one or more variables with coefficients in a number field form a *function field*. In the Galois theory of equations, we employ a field $R(k_1, \ldots, k_m)$ composed of all rational functions with rational coefficients of $k_1, \ldots, k_m$, either all of which are constants, or certain of which are constants and the others are given functions of specified variables not necessarily independent. For example, $R(3^{\frac{1}{2}})$ is composed of all numbers

$$\frac{a + b3^{\frac{1}{2}}}{c + d3^{\frac{1}{2}}},$$

in which a, b, c, d are rational numbers and c and d are not both zero. Multiplying numerator and denominator by $c - d3^{\frac{1}{2}}$, we see that $R(3^{\frac{1}{2}})$ is composed of all numbers $e + f3^{\frac{1}{2}}$ in which e and f are rational. Hence $R(3^{\frac{1}{2}}, x)$ is composed of all rational functions of x whose coefficients are of the form $e + f3^{\frac{1}{2}}$ with e and f rational.

The elements, whether numbers or functions, of a field are conveniently called *quantities*.

Although the set composed of the number zero alone satisfies the above definition of a number field, we shall exclude it and hence agree that every field F contains a quantity $q \neq 0$. Hence

§ 78] GREATEST COMMON DIVISOR 151

F contains $q/q = 1$, $1 + 1 = 2$, etc., and therefore contains all rational numbers. In other words, the field R of all rational numbers is a subfield of every field F. The foregoing field $R(k_1, \ldots, k_m)$ is said to be derived from R by the *adjunction* of $k_1, \ldots, k_m$ to R. If $n < m$, $R(k_1, \ldots, k_m)$ is derived from $R(k_1, \ldots, k_n)$ by the adjunction of $k_{n+1}, \ldots, k_m$.

EXERCISES

1. If $i^2 = -1$, every number of $R(i)$ can be expressed in the form $e + fi$, where e and f are rational numbers.
2. $R(3^{\frac{1}{2}}) = R(12^{\frac{1}{2}})$.
3. Write $r = 3^{\frac{1}{2}} i$, $\omega = -\frac{1}{2} + \frac{1}{2} r$. Then $R(r) = R(\omega) = R(\omega^2)$.
4. If $s = i + 2^{\frac{1}{2}}$, $R(i, 2^{\frac{1}{2}}) = R(s)$. Hint: $i - 2^{\frac{1}{2}} = -3/s$
5. In Ex. 4, $R(s)$ is derived from $R(i)$ by adjoining $2^{\frac{1}{2}}$.

78. Greatest common divisor. Let $f(x)$ and $g(x)$ be polynomials in x with coefficients in a field F, such that $g(x)$ is not identically zero. Let the division of $f(x)$ by $g(x)$ yield a quotient $a(x)$ and a remainder $r(x)$ whose degree is less than that of $g(x)$. Similarly, let the division of g by r yield a quotient b and a remainder s whose degree in x is less than that of r. Proceeding in this manner, let

$$f = ga + r, \quad g = rb + s, \quad r = sc + t, \quad s = td,$$

where the fourth remainder has been assumed to be identically zero, to simplify the notations. Since $a, r, b, \ldots$ were obtained by rational operations, all of their coefficients are quantities of F.

Any common divisor of f and g is seen in turn to be a divisor of r, s, and t. By employing the equations in reverse order, we see that t is a divisor of $s, r, g,$ and f. Since t is a divisor of f and g and since any common divisor of them is a divisor of t, we call t a *greatest common divisor* of f and g. It is determined uniquely apart from a factor belonging to F; in particular, there is a single greatest common divisor having unity as coefficient of the highest power of x.

The displayed equations give in turn

$$r = f - ga, \qquad s = g(1 + ab) - fb,$$
$$t = f(1 + bc) - g(a + c + abc).$$

THEOREM 1. *Any two polynomials $f(x)$ and $g(x)$, not both identically zero with coefficients in a field F, have a unique greatest common divisor $d(x)$ whose leading coefficient is unity. All of its coefficients belong to F, and there exist two polynomials $u(x)$ and $v(x)$ with coefficients in F such that*

$$d(x) \equiv u(x)f(x) + v(x)g(x).$$

The preceding discussion applies also when f and g denote integers. Let g be positive. Then the division of f by g yields an integral quotient a and an integral remainder r such that $0 \leq r < g$. Hence f and g have a unique positive greatest common divisor. In case it is unity, we call f and g *relatively prime*. Hence we have

THEOREM 2. *If f and g are relatively prime integers, there exist integers u and v such that $uf + vg = 1$.*

79. Roots of algebraic equations. Consider an equation

(1) $\qquad f(x) \equiv x^n + c_1 x^{n-1} + \cdots + c_n = 0.$

When $n = 3$ or 4, we found formulas in Ch. VII which express the n roots of (1) in terms of the coefficients c_i. But when $n > 4$, no such formulas are known except for special equations like $x^n - c = 0$.

§ 80] REDUCIBILITY

What shall be meant by roots of (1) when the coefficients c_i are independent complex variables? In that case the method (§78) of finding the greatest common divisor of $f(x)$ and its derivative $f'(x)$ leads to a final remainder $r(c_1, \ldots, c_n)$ which is free of x and is not zero identically. For, if $f(x) \equiv x^n - 1$, then $f'(x) = nx^{n-1}$ and $r = -1$. Let $C_1, \ldots, C_n$ be constant values of $c_1, \ldots, c_n$ such that $r(C_1, \ldots, C_n) \neq 0$. Then the resulting function (1) has no factor in common with its derivative, and equation (1) has n distinct roots. Since the polynomial r is a continuous function of the c's, we have $r(c_1, \ldots, c_n) \neq 0$ for all sets of values of $c_1, \ldots, c_n$ for which each difference $c_i - C_i$ is sufficiently small in absolute value. For each such set of values of $c_1, \ldots, c_n$, equation (1) has n distinct roots $X_1, \ldots, X_n$, and each X_i is known[1] to vary continuously when the c's vary continuously. These numbers X_i are therefore the values of a continuous function x_i of $c_1, \ldots, c_n$ for those values of the c's such that each difference $c_i - C_i$ is sufficiently small in absolute value. These functions $x_1, \ldots, x_n$ are called the roots of (1). In brief, the root values may be assembled into root functions.

If the c_i are not independent variables, but are functions of certain variables v_j such that $r(c_1, \ldots, c_n)$ is not zero identically in the v_j, the preceding discussion remains valid if we consider only sets of values of $c_1, \ldots, c_n$ which are obtained by varying the v_j continuously.

80. Reducibility. A polynomial in one variable x with coefficients in a field F is called *reducible in F* if it can be expressed as a product of polynomials in x, neither a constant, with coefficients in F. It is *irreducible in F* if no such factorization is possible.

For example, $x^2 - 4$ is reducible and $x^2 - 2$ is irreducible in the field R of rational numbers. But $x^2 - 2$ is reducible in $R(2^{\frac{1}{2}})$.

According as $f(x)$ is reducible or irreducible in F, the equation $f(x) = 0$ is called reducible or irreducible in F.

[1] Weber's Lehrbuch der Algebra, ed. 2, I, 1898, 148. Cf. Coolidge, Annals of Math., (2), 9, 1908, 116–8.

Theorem 3. *Let $f(x)$ and $g(x)$ be polynomials with coefficients in a field F and let $f(x)$ be irreducible in F. If one root α of $f(x) = 0$ satisfies $g(x) = 0$, then $f(x)$ is a divisor of $g(x)$.*

The greatest common divisor $d(x)$ of $f(x)$ and $g(x)$ is not a constant, since it has the factor $x - \alpha$. By §78, the coefficients of $d(x)$ belong to F. The quotient of $f(x)$ by $d(x)$ is a constant, since otherwise $f(x)$ would not be irreducible in F. Hence $d(x) = cf(x)$, where c is a quantity independent of x belonging to F. But $d(x)$ divides $g(x)$. Hence $f(x)$ divides $g(x)$.

Exercises

1. A root of an equation irreducible in F does not satisfy an equation of lower degree with coefficients in F.

2. An irreducible equation $f(x) = 0$ has no multiple root. Hint: Consider $f'(x) = 0$.

3. Two equations each irreducible in F are identical if they have a common root.

4. $x^4 + 1$ is reducible in any field which contains one of the four numbers $2^{\frac{1}{2}}$, $(-2)^{\frac{1}{2}}$, $i = (-1)^{\frac{1}{2}}$, $\rho = (1 + i)2^{-\frac{1}{2}}$, but is irreducible in all other fields. Hints: Its linear factors are $x \pm \rho$ and $x \pm \rho^{-1}$, while its pairs of quadratic factors are $x^2 \pm i$, $x^2 \pm 2^{\frac{1}{2}}x + 1$, $x^2 \pm 2^{\frac{1}{2}}ix - 1$.

81. Gauss's Lemma. *If a polynomial $f(x)$ with integral coefficients, that of the highest power of x being unity, is the product of two polynomials with rational coefficients,*

$$\phi(x) = x^m + b_1 x^{m-1} + \cdots + b_m,$$
$$\psi(x) = x^n + c_1 x^{n-1} + \cdots + c_n,$$

then these coefficients are all integers.

Let the fractions $b_1, \ldots, b_m$ be brought to their least positive common denominator B_0 and write $b_i = B_i/B_0$. Then B_0, $B_1, \ldots, B_m$ are integers having no common integral divisor >1.

§ 81]   GAUSS'S LEMMA   155

Similarly, let $c_i = C_i/C_0$, where $C_0, C_1, \ldots, C_n$ are integers having no common divisor > 1. The theorem is obviously true if $B_0 = 1$, $C_0 = 1$. Next, let $B_0 C_0 > 1$. Multiplying the members of $f \equiv \phi \psi$ by $B_0 C_0$, we get

(2) $$B_0 C_0 f(x) \equiv gh,$$

where
$$g = B_0 x^m + B_1 x^{m-1} + \cdots + B_m,$$
$$h = C_0 x^n + C_1 x^{n-1} + \cdots + C_n.$$

Let p be a prime factor of $B_0 C_0$. Since p divides each coefficient of the left member of (2), it divides each coefficient of gh. Not all coefficients of g are divisible by p. Let B_i be the first coefficient of g which is not divisible by p. Let C_k be the first coefficient of h which is not divisible by p. The total coefficient of $x^{m+n-i-k}$ in gh is

$$\cdots + B_{i+2} C_{k-2} + B_{i+1} C_{k-1} + B_i C_k$$
$$+ B_{i-1} C_{k+1} + B_{i-2} C_{k+2} + \cdots$$

Since $B_{i-1}, B_{i-2}, \ldots, B_0$ and $C_{k-1}, C_{k-2}, \ldots, C_0$ are all divisible by p, while $B_i C_k$ is not, the preceding sum is not divisible by p, contrary to the above.

Exercises

1. Gauss's Lemma for the case $m = 1$ shows that if an equation $x^r + a_1 x^{r-1} + \cdots + a_r = 0$ with integral coefficients, that of the highest power of x being unity, has a rational root, that root is an integer.

2. An integral root of the equation in Ex. 1 is an exact divisor of the constant term a_r. Hint: It divides all terms except the last.

3. Show that $x^3 - 2$, $x^3 - 3x + 1$, and $x^3 - 7x + 7$ are all irreducible in the field R of rational numbers. Hint: If such a cubic function were reducible in R, it would be the product of a linear and a quadratic function with coefficients in R. Apply the theorems in Exs. 1, 2.

82. Irreducibility of $x^p - A$.

Theorem 4. *If p is a prime and if A is a quantity of a field F such that A is not the pth power of any quantity of F, then $x^p - A$ is irreducible in F.*

An imaginary pth root of unity is given by

$$\rho = \cos 2\pi/p + i \sin 2\pi/p,$$

since $\rho^p = 1$ by De Moivre's theorem. By the same theorem, $\rho^l \neq 1$ if $0 < l < p$. Hence $1, \rho, \rho^2, \ldots, \rho^{p-1}$ are all distinct and give all the pth roots of unity. Thus if r denotes one root of $x^p = A$, all its roots are given by

(3) $\qquad\qquad r, \rho r, \rho^2 r, \ldots, \rho^{p-1} r.$

Suppose that $x^p - A$ has a factor $f(x)$ of degree $t < p$ with coefficients in F, that of x^t being unity. The product of the roots of $f(x) = 0$ is, apart from sign, equal to the constant term. Hence $\rho^s r^t$ belongs to F. By Theorem 2, there exist integers y and z such that $ty - pz = 1$. Hence F contains

$$(\rho^s r^t)^y = \rho^{sy} r^{pz+1} = \rho^{sy} r A^z = r' A^z,$$

where r' is one of the roots (3). Since A^z is in F, also r' is in F. In other words, A is the pth power of a quantity r' of F, contrary to hypothesis. This contradiction proves that $x^p - A$ is irreducible in F.

The theorem is illustrated by Ex. 3 of §81.

83. Irreducibility of the cyclotomic equation. If p is a prime, the equation

$$\frac{x^p - 1}{x - 1} \equiv x^{p-1} + x^{p-2} + \cdots + x + 1 = 0$$

has as its roots all the imaginary pth roots of unity. As a generalization, the equation

$$f(x) \equiv \frac{x^{tp} - 1}{x^t - 1} \equiv x^{t(p-1)} + x^{t(p-2)} + \cdots + x^t + 1 = 0 \qquad (t = p^{s-1})$$

has as its roots all the *primitive* p^sth roots of unity, viz., roots of $x^{p^s} = 1$ which do not satisfy $x^t = 1$.

Let ρ be an arbitrarily chosen primitive p^sth root of unity. Then $\rho^p, \rho^{2p}, \ldots, \rho^{tp} = 1$ give all the t roots of $x^t = 1$. The remaining $e = tp - t$ powers of ρ, with positive exponents $a_1, a_2, \ldots, a_e$ less than tp and not divisible by p, are roots of $f(x) = 0$ and give all its roots.

For example, if $p^s = 9$, then ρ^3, ρ^6, ρ^9 give the roots of $x^3 = 1$, while $\rho, \rho^2, \rho^4, \rho^5, \rho^7, \rho^8$ are the roots of $x^6 + x^3 + 1 = 0$ and are the primitive ninth roots of unity.

Suppose that $f(x)$ is reducible in the field of rational numbers. Then by Gauss's Lemma, $f(x)$ is the product of two polynomials $\phi(x)$ and $\psi(x)$, neither a constant, having integral coefficients, such that each has unity as the coefficient of the highest power of x. Since $f(1) = p$, we have $p = \phi(1)\psi(1)$. Since p is a prime, one of these integers, say $\phi(1)$, has the value ± 1. At least one of the roots $\rho^{a_1}, \ldots, \rho^{a_e}$ of $f(x) = 0$ is a root of $\phi(x) = 0$; hence

$$\phi(\rho^{a_1})\phi(\rho^{a_2}) \cdots \phi(\rho^{a_e}) = 0.$$

Here ρ is an arbitrarily chosen primitive p^sth root of unity and hence is any one of the roots of $f(x) = 0$. This proves that the function

$$P(x) \equiv \phi(x^{a_1})\phi(x^{a_2}) \cdots \phi(x^{a_e})$$

vanishes when x is replaced by any one of the roots of $f(x) = 0$ (which has no multiple root). Hence

$$P(x) \equiv f(x) \cdot q(x),$$

where the polynomial quotient $q(x)$ has integral coefficients since the leading coefficient of the divisor $f(x)$ is unity (or by use of Gauss's Lemma). Taking $x = 1$ and recalling that $f(1) = p$, we get

$$[\phi(1)]^e = (\pm 1)^e = p \cdot q(1),$$

where $q(1)$ is an integer. Since this is impossible, we have the following

THEOREM 5. *The cyclotomic equation $f(x) = 0$ whose roots are the primitive p^sth roots of unity, where p is a prime, is irreducible in the field R of rational numbers.*

EXERCISES

1. For $p^s = 3$, $f(x) = x^2 + x + 1$. For $p^s = 2^2$, $f(x) = x^2 + 1$. Verify that these quadratic functions are irreducible in R.

2. For $p^s = 5$, $f(x) = x^4 + x^3 + x^2 + x + 1$. Verify that it is irreducible in R by discussing $f(x) \equiv (x^2 + ax + r)(x^2 + bx + r^{-1})$.

Chapter IX

GROUP OF AN EQUATION FOR A GIVEN FIELD

84. Introduction. With a given algebraic equation we shall associate a certain group G of substitutions on its roots. The theory was initiated by E. Galois,[1] who was killed in a duel in 1832 at the age of 21. In a later chapter we shall prove that the equation is solvable by radicals if and only if G is a group of the kind called solvable and shall conclude that the general equation of degree five or more is not solvable by radicals. In another chapter we shall apply our theory to various questions ignored in elementary geometry and, for example, prove that it is not possible to trisect every angle with ruler and compasses.

85. Equal functions of the roots. Consider an equation

(1) $$f(x) \equiv x^n + c_1 x^{n-1} + \cdots + c_n = 0$$

whose coefficients are functions of certain variables v_j. In § 79 we explained the sense in which (1) has n roots which are functions of the v_j.

Two polynomials ϕ and ψ in the roots $x_1, \ldots, x_n$ are called *equal* if they have the same value for all sets of values of $c_1, \ldots, c_n$ employed there in defining the roots x_i. In case ϕ and ψ are rational functions of the x_i, we impose the further restriction on the c_i that the denominators of ϕ and ψ shall not be zero.

We shall call ϕ and ψ *distinct* if they are not equal.

For example, consider the reciprocal quartic equation

(2) $$x^4 + ax^3 + bx^2 + ax + 1 = 0,$$

[1] Jour. de Math., 11, 1846, 381–444; Œuvres, 1897; Manuscrits de Galois, 1908. For sources on Galois's life, see Bull. Amer. Math. Soc., 4, 1897–8, 332; Scientific Monthly, 1921, 363–75.

where a and b are independent complex variables. Since the reciprocal of any root is also a root, we may choose the notations of the roots so that $x_1 x_2 = 1$, $x_3 x_4 = 1$. Hence the two functions $x_1 x_2$ and $x_3 x_4$ are equal. Thus the rational functions x_1/x_3 and x_4/x_2 are equal. Again, for

$$t = x_1 + x_2 - x_3 - x_4, \qquad y_1 = x_1 x_2 + x_3 x_4,$$

we have

$$t^2 = (\sum x_1)^2 - 4 \sum x_1 x_2 + 4 y_1.$$

Hence the functions t^2 and $a^2 - 4b + 4y_1$ are equal.

If the function ψ is derived from ϕ by applying the substitution s on $x_1, \ldots, x_n$ and if ψ and ϕ are equal in the foregoing sense, we shall say that ϕ is *unaltered* by the substitution s.

For the quartic (2), $x_1 x_2$ is unaltered by the two substitutions (13)(24) and (1324), but is altered by (13).

86. Function of the n roots with n! values. In our study of an equation $f(x) = 0$ of degree n, we may assume without real loss of generality that the n roots are distinct. Otherwise, $f(x)$ and its derivative $f'(x)$ would have a greatest common divisor $g(x)$ not a constant, and $f(x)/g(x) = 0$ has no multiple root and is satisfied by every root of $f(x) = 0$. Hence let $f(x) = 0$ have the distinct roots $x_1, \ldots, x_n$. We shall prove

THEOREM 1. *There exist integers $m_1, \ldots, m_n$ such that*

(3) $$V_1 = m_1 x_1 + m_2 x_2 + \cdots + m_n x_n$$

gives rise to $n!$ distinct functions V_s when the $n!$ substitutions s on $x_1, \ldots, x_n$ are applied to V_1.

The subscript 1 indicates that we denote the identity substitution by 1 instead of I.

For $n = 2$, we may evidently take $V_1 = x_1$. The case $n = 3$ was discussed in Ex. 5, §67.

For any $n > 1$, we are to prove that we can choose integers $m_1, \ldots, m_n$ which satisfy no one of the $p = n!(n! - 1)/2$ equations $V_s = V_t$, where s and t are distinct substitutions. If s is the identity and $t = (x_1 x_2)$, the equation reduces to $m_1 = m_2$, and this is the only one of our p equations which involves only m_1 and m_2. It is not satisfied if $m_1 = 0$, $m_2 = 1$.

Next, let s and t leave $x_4, \ldots, x_n$ unaltered. If s and t replace x_1, x_2, x_3 by x_i, x_j, x_k and x_a, x_b, x_c, respectively, then $V_s = V_t$ becomes

$$m_1 x_i + m_2 x_j + m_3 x_k = m_1 x_a + m_2 x_b + m_3 x_c.$$

Equations of this type are the only ones of our p equations which involve only m_1, m_2, m_3. Since the case in which m_3 does not occur was treated before, we may assume that $k \neq c$. For $m_1 = 0$, $m_2 = 1$, the displayed equation reduces to

$$m_3 = \frac{x_b - x_j}{x_k - x_c}.$$

Hence we choose as m_3 any integer which is distinct from each of these fractions, finite in number.

Next, let s and t leave $x_5, \ldots, x_n$ unaltered, but replace x_4 by different roots. For $m_1 = 0$, $m_2 = 1$, and for a chosen integral value of m_3, $V_s = V_t$ expresses m_4 as a fractional function of the roots. Choose as m_4 any integer which is distinct from each of these fractions.

In the final step of the proof we employ substitutions s and t which alter x_n, and choose m_n as an integer distinct from each of a finite number of fractional functions of the roots.

Exercises

1. The roots of $x^3 + x^2 + x + 1 = 0$ are $x_1 = -1$, $x_2 = i$, $x_3 = -i$. Show that $x_2 + m_3 x_3$ is six-valued if m_3 is distinct from $0, 1, \pm i, 1 \pm i$, $\frac{1}{2}(1 \pm i)$, and hence if $m_3 = -1$. Thus also $x_2 - x_1$ is six-valued.

Denote the six substitutions on x_1, x_2, x_3 by

$$I, a = (12), b = (13), c = (23), d = (123), e = (132).$$

Compute the following functions and verify that they are distinct:

$$V_1 = x_2 - x_1 = 1 + i, \quad V_b = x_2 - x_3 = 2i, \quad V_c = x_3 - x_1 = 1 - i,$$
$$V_a = -V_1, \quad V_d = -V_b, \quad V_e = -V_c.$$

2. If $x_1, \ldots, x_n$ are independent variables, the function (3) has $n!$ distinct values if $m_1, \ldots, m_n$ are any distinct constants.

87. Galois resolvents. Let $s_1, \ldots, s_r$ denote the substitutions on $x_1, \ldots, x_n$, where $r = n!$ and s_1 is the identity substitution. Let V_1 be the r-valued function (3). If we apply s_j to V_1 and then apply s_k to the resulting function V_{s_j}, we obtain V_{s_l}, where $s_l = s_j s_k$. When k is fixed, but j takes the values $1, 2, \ldots, r$, then l takes the same values in some new order. Hence s_k merely permutes $V_1, \ldots, V_{s_r}$ amongst themselves. The elementary symmetric functions of these V's are therefore symmetric functions of $x_1, \ldots, x_n$, and hence are polynomials in $c_1, \ldots, c_n$ with integral coefficients. Let F be any field containing $c_1, \ldots, c_n$. Thus the coefficients of the polynomial defined by the expansion of

(4) $$P(V) \equiv (V - V_1)(V - V_{s_2}) \cdots (V - V_{s_r})$$

are quantities in F.

If $P(V)$ is reducible in F, let $G(V)$ be that factor irreducible in F for which $G(V_1) = 0$. But if $P(V)$ is irreducible in F, take $G(V)$ to be $P(V)$ itself. In either case, $G(V) = 0$ is an equation with the root V_1 whose coefficients belong to F and which is irreducible in F; it is called a *Galois resolvent* of equation (1) for the field F.

The corresponding resolvent of the equation in Ex. 1 of §86 for the field of rational numbers is evidently

$$G(V) \equiv (V - V_1)(V - V_c) \equiv V^2 - 2V + 2 = 0;$$

but for the field $R(i)$, the resolvent is $V - V_1 = 0$.

§ 87] GALOIS RESOLVENTS 163

THEOREM 2. *Let $\phi(x_1, \ldots, x_n)$ be any polynomial, with coefficients in a field F, in the roots x_i of an equation with coefficients in F. Let s be any substitution on the roots and let it replace ϕ by ϕ_s and V_1 by V_s, where V_1 is the $n!$-valued function (3) with integral coefficients. Then*

$$(5) \qquad \phi_s = \frac{\lambda(V_s)}{P'(V_s)},$$

where λ is a polynomial with coefficients in F, while P' is the derivative of the polynomial (4) *whose coefficients belong to F, whence $P'(V_s) \neq 0$. Thus ϕ_s is the same rational function $\rho(V_s)$ of V_s that $\phi \equiv \phi_1$ is of V_1.*

Let $r = n!$, $l = s_r$. Evidently s_k permutes $\phi_1, \ldots, \phi_l$, in the same manner that it permutes $V_1, \ldots, V_l$. Hence the fractions in

$$(6) \qquad \lambda(V) \equiv \phi_1 \frac{P(V)}{V - V_1} + \cdots + \phi_l \frac{P(V)}{V - V_l}$$

are merely permuted amongst themselves by any substitution on $x_1, \ldots, x_n$. Replacing $P(V)$ by the product (4), we see that $\lambda(V)$ becomes a polynomial in V whose coefficients are integral rational symmetric functions of $x_1, \ldots, x_n$ with coefficients in F, and hence are equal to quantities in F. For $V = V_s$ each fraction in (6) becomes zero except that with the denominator $V - V_s$. If in the quotient of the product (4) by $V - V_s$ we take $V = V_s$, we get the value of the derivative of (4) for $V = V_s$. Hence

$$\lambda(V_s) = \phi_s P'(V_s).$$

In case the degree of the Galois resolvent $G(V) = 0$ is less than $n!$, we may express the numerator and denominator of $\phi_1 = \lambda(V_1)/P'(V_1)$ as polynomials of lower degree in V_1 by means of $G(V_1) = 0$. Let $r(V_1)$ be the resulting fractional expression for ϕ_1. It does not now follow (as in the last part of the theorem) that we can apply any substitution s to the members of the relation $\phi_1 = r(V_1)$ and obtain a correct relation $\phi_s = r(V_s)$. To obtain the latter we would need to reduce the fraction (5) by means of $G(V_s) = 0$, which however is not known to hold unless s is one of the special substitutions for which V_s is a root of $G(V) = 0$. For the remaining substitutions s, V_s is a root of a factor, other than $G(V)$, of $P(V)$ in (4). This discussion brings us to a critical point in our theory and indicates the care which must be taken in its development.

88. The group of an equation for a given field. Let F be a field containing the coefficients of an equation (1) whose roots $x_1, \ldots, x_n$ are distinct. Let the roots of a Galois resolvent $G(V) = 0$ of degree g be

(7) $$V_1, V_a, V_b, \ldots, V_p,$$

in which the subscripts denote the substitutions on $x_1, \ldots, x_n$ by which these V's are derived from V_1.

THEOREM 3. *These g substitutions*

(8) $$1, a, b, \ldots, p$$

form a group G, called the group of the given equation (1) *for the field F.*

We are to prove that the product rs of any two equal or distinct substitutions (8) is one of those substitutions. Take V_r as the function ϕ in Theorem 2. Then

$$V_r = \frac{\lambda(V_1)}{P'(V_1)}, \quad V_{rs} = (V_r)_s = \frac{\lambda(V_s)}{P'(V_s)},$$

where $\lambda(V)$ and $P'(V)$ are polynomials in V with coefficients in F. Since V_r is a root of $G(V) = 0$, the equation

$$G\left(\frac{\lambda(V)}{P'(V)}\right) = 0$$

is satisfied when $V = V_1$. The product of the left member by the gth power of $P'(V)$ is a polynomial $H(V)$ in V which vanishes for $V = V_1$ and has quantities in F as coefficients. Hence (§80) the root V_s of the equation $G(V) = 0$ irreducible in F is a root of $H(V) = 0$. Since $P'(V_s) \neq 0$, we may divide $H(V_s)$ by the gth power of $P'(V_s)$; we get

$$0 = G\Big(\frac{\lambda(V_s)}{P'(V_s)}\Big) = G(V_{rs}).$$

Hence V_{rs} is one of the functions (7), so that rs is one of the substitutions (8).

EXAMPLE. For the field R of rational numbers, a Galois resolvent of $x^3 + x^2 + x + 1 = 0$ was seen in §87 to have the two roots V_1 and V_c. Hence the group of this cubic for R is $\{1, (x_2\,x_3)\}$. But for the field $R(i)$, a Galois resolvent is $V - V_1 = 0$ and the group is the identity.

89. Characteristic properties A and B of the group G of a given equation for a given field F.

A. *If a rational function with coefficients in F of the roots of an equation with coefficients in F remains unaltered in value by all of the substitutions of the group G of the equation for F, it is equal to a quantity in F.*

Let ϕ/ψ be a quotient of two polynomials in the roots with coefficients in F. The coefficients of the equation are of course restricted to values for which $\psi \neq 0$. We have formula (5) and similarly

$$\psi_s = \frac{\mu(V_s)}{P'(V_s)},$$

where $\mu(V)$ is a polynomial in V with coefficients in F. Since $\psi_1 \neq 0$, $\mu(V_1) \neq 0$. If s is any substitution of G, then $\psi_s \neq 0$. For, if $\mu(V) = 0$ has the root V_s, it has also the root V_1 in common with the irreducible Galois resolvent $G(V) = 0$ (§80). Hence the functions

$$\frac{\phi_s}{\psi_s} = \frac{\lambda(V_s)}{\mu(V_s)} \qquad (s = 1, a, b, \ldots, p)$$

are defined for each substitution s of the group G.

Suppose that these g functions are equal, or in other words that ϕ/ψ is unaltered by all substitutions of G. Then

$$\frac{\phi}{\psi} = \frac{1}{g}\left\{\frac{\lambda(V_1)}{\mu(V_1)} + \frac{\lambda(V_a)}{\mu(V_a)} + \cdots + \frac{\lambda(V_p)}{\mu(V_p)}\right\}.$$

The second member is a rational symmetric function with coefficients in F of the roots (7) of $G(V) = 0$ and hence is equal to a rational function of its coefficients, which belong to F. Hence ϕ/ψ is equal to a quantity in F.

B. *Conversely, if a rational function of the roots with coefficients in F is equal to a quantity in F, it remains unaltered in value by all of the substitutions of G.*

Let $\phi/\psi = r$, where r is in F. Then $\lambda(V)/\mu(V) - r$ vanishes for $V = V_1$. Hence the equation $\lambda(V) - r\mu(V) = 0$ with coefficients in F is satisfied by every root V_s of the irreducible equation $G(V) = 0$ (§80). Hence

$$r = \frac{\lambda(V_s)}{\mu(V_s)} = \frac{\phi_s}{\psi_s} \qquad (s = 1, a, b, \ldots, p),$$

so that ϕ/ψ is unaltered by all the substitutions of G.

Since an $n!$-valued function V_1 with coefficients in a given field F can be chosen in an infinitude of ways, there are infinitely many Galois resolvents $G(V) = 0$. Our definition of the group G of the given equation for the field F was based upon a single such resolvent, which is determined by the special V_1 chosen. But different functions V_1 lead to the same group. In fact, G is uniquely determined by the given equation and field F. This is a consequence of the following

THEOREM 4. *The group of a given equation for a given field F is uniquely defined by properties* A *and* B.

First, suppose that $H = \{1, r, \ldots, m\}$ is a group for which property A holds. Then the coefficients of

$$\phi(V) \equiv (V - V_1)(V - V_r) \cdots (V - V_m),$$

being symmetric functions of $V_1, V_r, \ldots, V_m$, are unaltered by the substitutions of H and hence are equal to quantities in F. Since the equation $\phi(V) = 0$, with coefficients in F, admits one root V_1 of the irreducible Galois resolvent $G(V) = 0$, it admits all of the roots (7) of the latter (§80). Hence $1, a, \ldots, p$ occur among the substitutions of H, so that G is a subgroup of H.

Second, suppose that $K = \{1, p, \ldots, t\}$ is a group for which property B holds. Then the Galois function $G(V_1)$, being equal to the number zero of F, remains unaltered in value by the substitutions of K, so that

$$0 = G(V_1) = G(V_p) = \cdots = G(V_t).$$

Hence $V_p, \ldots, V_t$ occur among the roots (7) of $G(V) = 0$. Thus K is a subgroup of G.

If $H = K$ is a group for which both properties A and B hold, the two results show that this group coincides with G.

In view of its repeated application below, we state the result obtained in the second case as the

THEOREM 5. *If every rational function of the roots with coefficients in F which is equal to a quantity in F is unaltered in value by every substitution of a group K, then K is a subgroup of the group G for F of the equation.*

90. Transitive and regular groups. A group of substitutions on $x_1, \ldots, x_n$ is called *transitive* if it contains substitutions s_i which replace x_1 by x_i for $i = 1, \ldots, n$. Then it contains a substitution $s_i^{-1} s_j$ which replaces x_i by x_j. A group which is not transitive is called *intransitive*. A transitive group of order n on n letters is called *regular*.

The symmetric group on n letters and the group

(9) $\qquad G_4 = \{I, \ (12)(34), \ (13)(24), \ (14)(23)\}$

on four letters are both transitive. The latter is regular, while the former is not if $n > 2$. But $\{I, \ (12), \ (34), \ (12)(34)\}$ is an intransitive group.

Theorem 6. *The order of any transitive group G on n letters is divisible by n.*

All those substitutions of G which leave x_1 unaltered form a subgroup H. Since G is transitive, it contains a substitution s_i which replaces x_1 by x_i. Hence

$$G = H + Hs_2 + Hs_3 + \cdots + Hs_n,$$

where every substitution of the set Hs_i replaces x_1 by x_i and hence is distinct from each substitution of any set Hs_j having $j \neq i$. Next, every substitution t of G is in one of the sets Hs_i. For, if t replaces x_1 by x_k, then ts_k^{-1} leaves x_1 unaltered and hence is a substitution h of H, whence $t = hs_k$. Hence the order of G is the product of the order by H by n.

Theorem 7. *If an equation is irreducible in a field F, its group for F is transitive, and conversely.*

Let $f(x) = 0$ be irreducible in F. Contrary to the theorem, suppose that its group G for F is intransitive and contains substitutions replacing x_1 by $x_1, x_2, \ldots, x_m$, but none replacing x_1 by one of $x_{m+1}, \ldots, x_n$. Consider any substitution s of G and let it replace x_i ($i \leq m$) by x_j. Since G contains a substitution t which replaces x_1 by x_i, it contains ts which replaces x_1 by x_j. Hence $j \leq m$, so that s permutes $x_1, \ldots, x_m$ amongst themselves, and leaves unaltered any symmetric function of them. Hence property A of §89 shows that the coefficients of the polynomial in x given by the expansion of

$$g(x) \equiv (x - x_1)(x - x_2) \cdots (x - x_m)$$

are quantities of F. Thus $f(x)$ has the factor $g(x)$ in F, contrary to its irreducibility in F.

To prove the converse of the theorem, let G be transitive and suppose that $f(x)$ is reducible in F. Then $f(x)$ has a factor $g(x)$ of degree $m < n$ with coefficients in F, such that $g(x) = 0$ has the root x_1. Since $g(x_1)$ is equal to the number zero of F, property B shows that it is unaltered in value by every substitution of G. Since G is transitive, it contains a substitution replacing x_1 by any chosen x_i ($i \leq n$). Hence $g(x_i) = 0$, in contradiction with $m < n$.

EXAMPLE. Find the group G of $x^3 - 7x + 7 = 0$ for the field R of rational numbers.

Since the equation is irreducible in R (Ex. 3, §81), G is transitive. By §67, the square of

(10) $\psi = (x_1 - x_2)(x_2 - x_3)(x_3 - x_1)$

is the discriminant 49 of our cubic equation, so that ψ is equal to a number ± 7 of R. Since any transposition replaces ψ by $-\psi$ and hence alters ψ in value, it is not in G by property B. The two results show that $G = \{I, (123), (132)\}$.

EXERCISES

1. For any field F containing the coefficients of an equation having a rational root x_1 and no multiple root, each substitution of its group leaves x_1 unaltered (by property B). If a root x_j is not in F, the group contains a substitution which alters x_j. Hence prove that the group of $(x - 1)(x + 1)(x - 2) = 0$ for the field R of rational numbers is the identity. But the group of $x^3 - 1 = 0$ for R is of order 2.

2. The group of $x^3 - 9x + 9 = 0$ for R is of order 3. Hint: The value of (10) is here ± 27.

3. The group of $x^3 - 2 = 0$ for $R(\omega)$, where ω is an imaginary cube root of unity, is of order 3. Use (10).

4. The group of $x^3 - 2 = 0$ for R is the symmetric group. For, if it were the alternating group, (10) would have a rational value.

5. Find the group G of $x^4 + 1 = 0$ for R.
Hints: By Ex. 1, §71, $y_1 = x_1 x_2 + x_3 x_4$, y_2, and y_3 are the distinct roots 0, ± 2 of $y^3 - 4y = 0$. Property B shows that each substitution s of G leaves $y_1, y_2,$ and y_3 formally unaltered. By Ex. 4, §73, s belongs to the group (9). Since $x^4 + 1$ is irreducible in R by Ex. 4, §80, G is transitive and hence of order ≥ 4. Thus $G = G_4$.

6. The group of $x^4 + 1 = 0$ for $R(i)$ is of order 2.
Hint: The factors $x^2 \pm i$ show that the group is an intransitive subgroup of (9).

7. If $a_0 x^n + a_1 x^{n-1} \cdots + a_n = 0$ has rational coefficients and is irreducible in R, then

(a) If there is a complex root of absolute value unity, the equation is a reciprocal equation of even degree.

(b) If there is a root $r + si$, where r is rational, then n is even and the n roots may be paired so that the sum of the two of any pair is $2r$, whence $r = - a_1/(n\, a_0)$. In particular, if $r = 0$, the equation involves only even powers of x.

(c) If there is an imaginary root $a + bi$ such that $a^2 + b^2$ is rational, then n is even and the n roots may be paired so that the product of any two of a pair is $a^2 + b^2$. If ρ is a square root of $a^2 + b^2$, then $\rho^n = a_n/a_0$. Replacing x by ρy, we obtain a reciprocal equation in y.

8. The group for R of $x^4 + x^3 + x^2 + x + 1 = 0$ is a cyclic group of order 4. It is irreducible in R by Ex. 2, §83.

9. Let G be the group for F of an equation irreducible in F whose n roots are rational functions $f_i(r)$ of one root r with coefficients in F for $i = 0, 1, \ldots, n - 1$. Prove that G is a regular group and that $f_i[f_j(r)]$ is a root.

91. Rational functions belonging to a group. Consider a rational function $\psi(x_1, \ldots, x_n)$, with coefficients in a field F, of the roots of an equation with coefficients in F. Let G be the group for F of the equation. Let $1, a, b, \ldots, k$ be all those substitutions of G which leave ψ unaltered in value. Since $\psi - \psi_a$

§91] FUNCTIONS BELONGING TO A GROUP 171

has the value zero, we have $\psi_b - \psi_{ab} = 0$ by property B. Hence $\psi_{ab} = \psi$, so that ab is in the set $1, a, \ldots, k$, which therefore forms a group H. We shall say that ψ *belongs* to the subgroup H of G.

EXAMPLE. Let x_1 and $x_2 = -x_1$ be two roots of $x^4 + 1 = 0$. The only substitutions of its group (9) for R which leave x_1^2 unaltered are the identity and (12)(34). Hence they form the subgroup to which x_1^2 belongs.

Conversely, let H be any given subgroup of G. Let V_1 be any $n!$-valued function of the roots with coefficients in F, and let $V_1, V_a, \ldots, V_k$ be the functions obtained from V_1 by applying the substitutions of H. We may choose an integer r such that

$$\psi \equiv (r - V_1)(r - V_a) \cdots (r - V_k)$$

belongs to H. For, if s is any substitution of H, then $s, as, \ldots, ks$ are distinct and are in H, and hence form a permutation of $1, a, \ldots, k$. Thus ψ is equal to

$$\psi_s \equiv (r - V_s)(r - V_{as}) \cdots (r - V_{ks}).$$

But if s is a substitution of G which is not in H, then ψ_s is not identical with ψ for all values of r, since V_s is different from $V_1, V_a, \ldots, V_k$. As in §86, we may choose an integer r such that every such ψ_s is distinct from ψ. Then ψ belongs to H. This proves

THEOREM 8. *Every rational function ψ with coefficients in F of the roots of an equation having the group G for the field F belongs to a definite subgroup of G. There exist such functions ψ belonging to any assigned subgroup of G.*

We next prove the important supplementary

THEOREM 9. *If a rational function ψ with coefficients in F of the roots of an equation having the group G for F belongs to a subgroup H of index u under G, then the substitutions of G replace ψ by exactly*

u distinct functions, called the **conjugates** to ψ under G. They are the roots of an equation with coefficients in F which is irreducible in F.

As in §76, let

(11) $\qquad G = H + Hg_2 + Hg_3 + \cdots + Hg_u.$

Let h be any substitution of H. Then

$$\psi_{hg_i} = (\psi_h)_{g_i} = \psi_{g_i} \equiv \psi_i,$$

so that ψ takes at most u values under G. If $\psi_i = \psi_j$ for $j < i$, then $\psi_p = \psi$, where $p = g_i g_j^{-1}$ is a substitution of G leaving ψ unaltered and hence is in H. Then $g_i = pg_j$, contrary to (11). Thus $\psi_1, \ldots, \psi_u$ are distinct.

Any substitution s of G merely permutes $\psi_1, \ldots, \psi_u$ amongst themselves. For, $g_i s$ is in G and hence by (11) may be expressed in the form hg_k, where h is in H; then

$$(\psi_i)_s = \psi_{g_i s} = \psi_{hg_k} = \psi_{g_k} = \psi_k.$$

Hence s leaves unaltered their elementary symmetric functions which are, apart from sign, the coefficients of the polynomial

(12) $\qquad g(y) \equiv (y - \psi_1)(y - \psi_2) \cdots (y - \psi_u).$

Therefore these coefficients are quantities in F by property A.

If $g(y)$ is reducible in F, it has a factor $t(y)$ with coefficients in F which is zero for $y = \psi_1$. By property B, $t(\psi_2) = 0, \ldots,$ $t(\psi_u) = 0$. Thus $t \equiv g$, so that $g(y)$ is irreducible.

EXAMPLE 1. The group G of $x^3 + x^2 + x + 1 = 0$ for R is $\{I, (23)\}$ if $x_1 = -1$. The conjugates to $\psi = x_2 - x_1$ under G are $\psi_1 = \psi$ and $\psi_2 = x_3 - x_1$; they are the roots of $y^2 - 2y + 2 = 0$, which is irreducible in R.

§ 91]   FUNCTIONS BELONGING TO A GROUP   173

EXAMPLE 2. The group G of $x^4 + 1 = 0$ for $R(i)$ is $\{I, s\}$, where s is a product of two transpositions. See Ex. 6, §90. Write $x_1 = \rho$, $x_2 = i\rho$, $x_3 = -\rho$, $x_4 = -i\rho$, where $\rho = (1+i)2^{-\frac{1}{2}}$, whence $\rho^2 = i$. Then $s = (13)(24)$. The conjugates x_1 and x_3 to x_1 under G are the roots of $y^2 - i = 0$, which is irreducible in $R(i)$ since this field does not contain ρ.

THEOREM[1] 10. *Let ϕ be a rational function with coefficients in F of the roots of an equation whose group for F is G. If ϕ remains unaltered by all those substitutions of G which leave unaltered another rational function ψ of the roots with coefficients in F, then ϕ is equal to a rational function of ψ with coefficients in F.*

Let H be the subgroup of G of index u to which ψ belongs. By means of (11), we obtain the u distinct conjugates $\psi_1, \ldots, \psi_u$ to $\psi = \psi_1$ under G. Since every substitution h of H leaves ϕ unaltered, each product hg_i replaces ϕ by $\phi_i \equiv \phi_{g_i}$. By the proof of Theorem 9, any substitution s of G replaces ψ_i by a certain ψ_k, and likewise ϕ_i by ϕ_k. Thus, if $g(y)$ is defined by (12), the fractions in

$$\lambda(y) \equiv g(y) \left(\frac{\phi_1}{y - \psi_1} + \frac{\phi_2}{y - \psi_2} + \cdots + \frac{\phi_u}{y - \psi_u} \right)$$

are merely permuted by s. The second member is equal to a polynomial $\lambda(y)$ each of whose coefficients is therefore unaltered by every substitution s of G and hence is equal to a quantity in F. Taking $\psi_1 \equiv \psi$ as y, we get (as in the proof of Theorem 2)

$$\phi = \lambda(\psi)/g'(\psi).$$

EXAMPLE. Let ψ be an $n!$-valued function V_1, so that the identity is the only substitution leaving ψ unaltered. Since it leaves every ϕ unaltered, Theorem 10 states that every rational function ϕ with coefficients in F is expressible as a rational function of V_1 with coefficients in F. This follows from the like property in §87 of the polynomial numerator and denominator of ϕ.

---

[1] The case in which the roots are independent variables is due to Lagrange. Nouv. Mém. Acad. Berlin, années 1770–1, §§ 100–4; Œuvres, III, 374–88.

92. Effect on the group by an adjunction to the field. Let G be the group of $f(x) = 0$ for a field F containing the coefficients. The field $F' = F(\psi)$ composed of the rational functions of ψ with coefficients in F is said to be derived from F by adjoining ψ to F (§77). If the irreducible Galois resolvent $G(V) = 0$ for the field F remains irreducible in F', the group for F' of $f(x) = 0$ is evidently G. But if it reduces in F', let $G'(V)$ be that factor of $G(V)$ which has its coefficients in F', is irreducible in F', and vanishes for $V = V_1$. Then, if $V_1, V_a, \ldots, V_k$ are the roots of $G'(V) = 0$, the group for F' of $f(x) = 0$ is by definition $G' = \{I, a, \ldots, k\}$, which is a subgroup of G. The former result may be combined with this since a group is included among its subgroups. Hence we have

THEOREM 11. *By an adjunction to the field, the group of an equation is reduced to a subgroup.*

EXAMPLE 1. For the field R of rational numbers the group of $x^3 + x^2 + x + 1 = 0$ is $G = \{I, (23)\}$ if x_1 denotes the root -1; but in the enlarged field $R(i)$, the group is the identity (Example in §88). Note that the adjoined quantity $i = x_2$ belongs to the identity subgroup of G.

EXAMPLE 2. The group for R of $x^4 + 1 = 0$ is G_4, given by (9). In the notations of Ex. 2, §91, the group for $R(i)$ is $G_2 = \{I, (13)(24)\}$. The latter is the subgroup of G_4 to which $i = x_1^2 = x_3^2 = -x_2^2 = -x_4^2$ belongs.

These examples illustrate also the important

THEOREM 12. *Let G be the group for F of an equation with coefficients in F. Let H be the subgroup to which belongs a rational function ψ of the roots with coefficients in F. By the adjunction of ψ to F, the group of the equation is reduced from G to H.*

It is to be proved that H has the characteristic properties A and B of the group of the equation for the field $F' = F(\psi)$. Evidently any rational function ϕ of the roots with coefficients in F' is equal to a rational function ϕ_1 of the roots with coefficients in F.

§ 92]   EFFECT OF AN ADJUNCTION   175

First, let ϕ be unaltered by all of the substitutions of H. By Theorem 10, ϕ_1 (which is unaltered by H) is a rational function of ψ with coefficients in F. Hence $\phi_1 = \phi$ is in $F' = F(\psi)$, so that property A holds for H and F'.

Second, let ϕ be equal to a quantity r in F', namely a rational function $r(\psi)$ of ψ with coefficients in F. Then $\phi_1 - r(\psi)$ is a rational function of the roots with coefficients in F having the value zero, and hence is unaltered by every substitution of G and, in particular, by the substitutions of H. The latter leave $r(\psi)$ unaltered and therefore also $\phi_1 = \phi$. This proves property B for H and F'.

EXERCISES

1. By the adjunction of $2^{\frac{1}{2}}$, the group G_2 of $x^4 + 1 = 0$ for $R(i)$ is reduced to the identity group.

2. By the adjunction of an imaginary cube root ω of unity, the group G_6 for R of $x^3 - 2 = 0$ is reduced to the cyclic group C_3 (Exs. 3, 4, §90). Verify that $\omega = x_2/x_1$ belongs to the group C_3. By the further adjunction of $2^{\frac{1}{3}}$, the group is reduced to the identity.

3. Find the group of $x^4 + x^3 + x^2 + x + 1 = 0$ for $R(5^{\frac{1}{2}})$.

93. Group of the general equation. Let the coefficients $c_1, \ldots, c_n$ of (1) be independent complex variables. Let $x_1, \ldots, x_n$ be the roots in the sense of §79.

THEOREM 13. *The group of the general equation for the field F determined by its coefficients and any constants finite in number is the symmetric group.*

Suppose the group is $H = \{1, a, \ldots, k\}$, which does not contain the substitution s. Apart from sign, the elementary symmetric functions of $V_1, V_a, \ldots, V_k$ are coefficients of the Galois resolvent and hence are equal to quantities in F. Hence if r is in F,

$$\psi \equiv (r - V_1)(r - V_a) \cdots (r - V_k)$$

is equal to a quantity in F. As in §91, we may choose an integral value of r such that ψ_s is distinct from ψ. But ψ is in F and hence is expressible rationally in terms of the symmetric functions $c_1, \ldots, c_n$ of the x's and is therefore unaltered by s. This contradiction proves the theorem.

94. Further results. By an elaborate analysis, Hilbert[1] proved that if f is a polynomial in $x_1, \ldots, x_r, t_1, \ldots, t_s$ with integral coefficients which is irreducible in R, it is always possible in an infinity of ways to assign integral values to the t_j such that f becomes a function of the x_i irreducible in R. Let $\phi(x) = 0$ be an equation in x whose coefficients are polynomials in $t_1, \ldots, t_s$ with integral coefficients. We can assign integral values to the t_j such that the resulting equation in x has the same group for R as $\phi(x) = 0$. Applying this to the general equation, we conclude the existence of an infinitude of equations of degree n with integral coefficients whose group for R is the symmetric group. Similarly for the alternating group. The last two results have been proved more simply by various writers.[2]

A beginning[3] has been made on the problem to find all equations of degree n with a prescribed group, with explicit results for $n = 3$ and $n = 4$.

An equation whose coefficients are rational functions of a complex variable k has a group which contains as an invariant subgroup the *monodromie* group composed of the substitutions on the roots which arise when k describes a closed path in the complex

---

[1] Jour. für Math., 110, 1892, 104–29. In Öfversigt Finska Vetenskaps-Soc. Förhandlingar, 59, 1916-7, No. 12, Wäisälä proved that integral values may be assigned to the t_j such that f decomposes into exactly as many irreducible factors as when the t_j are variables.

[2] Weber, Math. Papers Chicago Congress, 1896, 401–7; Algebra, ed. 2, I, 1898, 653–5. Maillet, Jour. de Math., (5), 5, 1899, 205–16. Bauer, Jour. für Math., 132, 1907, 33–35; Math. Annalen, 64, 1907, 325–7. Schur, Jahresber. Deutsch. Math. Vereinigung, 29, 1920, 145–50. Furtwängler, Math. Annalen, 85, 1922, 34–40.

[3] E. Noether, Math. Annalen, 78, 1918, 221–9. Seidelmann, *ibid.*, 230–3. Breuer, *ibid.*, 86, 1922, 108–13 (cyclic sextics).

plane.[4] Similarly there is a monodromie group of a linear differential equation obtained by using linearly independent integrals (instead of roots as before).

The group of a system of algebraic equations has been considered.[5] The groups of a reciprocal quartic[6] and De Moivre's[7] solvable quintic[8] have been found.

The group of the equation for the nine abscissas of the points of inflexion of a plane cubic curve has been treated fully by the author.[9] A more elaborate discussion is required for the 27 straight lines on a cubic surface or the 28 bitangents to a plane quartic curve.[10]

The Galois theory has been extended to ordinary linear differential equations[11] and to complete systems of linear partial differential equations.[12]

---

[4] Miller, Blichfeldt, and Dickson, Finite Groups, 378–81. Hermite, Comptes Rendus Paris, 32, 1851 (Œuvres, I, 276–80). Frobenius, Jour. für Math., 74, 1872, 254–72. Kneser, Irreducibilität und Monodromiegruppe alg. Gl., Thesis, Berlin, 1885; Math. Annalen, 28, 1886, 125–32. Hurwitz, ibid., 39, 1891, 23. Bianchi, Teoria dei Gruppi di Sostituzioni, 1897, 314–33. Baker, Annals of Math., 14, 1912–3, 119–36. Ritt, Trans. Amer. Math. Soc., 24, 1922, 21–30.

[5] Weisner, Bull. Amer. Math. Soc., 30, 1924, 314–6 (concept due to König, Math. Annalen, 18, 1881, 69–77).

[6] Dickson and Börger, Amer. Math. Monthly, 15, 1908, 71–78, 85–87. Miller, Blichfeldt, and Dickson, Finite Groups, 1916, 291. Criteria for irreducibility of a reciprocal equation of degree $2n$, Dickson, Bull. Amer. Math. Soc., 14, 1907–8, 426–30.

[7] Phil. Trans. London, 25, 1707, 2368–70.

[8] Börger, Amer. Math. Monthly, 15, 1908, 171–4.

[9] Miller, Blichfeldt, and Dickson, Finite Groups, 1916, 327–42; Annals of Math., 16, 1914, 50–66.

[10] Ibid., 343–75. On pp. 375–7 is a brief account of the groups of further geometrical problems.

[11] Picard, Traité d'Analyse, III, Ch. 17. Loewy, H. Weber Festschrift, 1912, 198–227.

[12] Vessiot, Annales Sc. École Normale Sup., (3), 21, 1904, 1–85.

Chapter X

EQUATIONS SOLVABLE BY RADICALS

95. Historical note. Euler, the leading mathematician of the eighteenth century, believed that every algebraic equation is solvable by radicals. Beginning in 1799, Ruffini published several attempts to prove that the general equation of degree > 4 is not solvable by radicals, employing substitutions on the roots. His final and simplest proof [1] is essentially the same as that now known as Wantzel's simplification [2] of Abel's proof; but he gave only a trivial remark by way of proof of the auxiliary theorem, now known as Abel's, that, if an equation is solvable by radicals, the roots can be given a form such that all the radicals occurring in their expressions are rational functions of the roots [and roots of unity, with rational coefficients].

Abel's [3] proof of the last theorem has two defects. As noted by Sir W. R. Hamilton,[4] the nth root of $q_\mu^n p^\mu$ is in general of order higher than $p^{1/n}$. This first defect is best remedied by ignoring Abel's unnecessary classification of radicals. Second, Abel's proof is vitiated by the occurrence of roots of unity and is remedied by adjoining to the field at the outset all kth roots of unity, where k takes the values of the indices of all the radicals involved in the expression for a root y.

But to give a sound proof of the impossibility of solving the general equation of degree > 4 along the lines of Ruffini and Abel, we must introduce the ideas of groups of substitutions and adjunc-

---

[1] Riflessioni intorno alla soluzione delle equazioni algebraiche generali, Modena, 1813.

[2] Serret, Algèbre, II, eds. 4 or 5, 512.

[3] Jour. für Math., 1, 1826, 65–84; Œuvres complètes, I, 1881, 66–94.

[4] Trans. Royal Irish Acad., 18, 1839, 171–259. A very complicated reconstruction of Abel's proof.

tions to fields, which served Galois in his far more general theory of the solvability of any equation.

96. Solvability by radicals. The four rational operations are addition, subtraction, multiplication, and division.

An algebraic equation is said to be solvable by radicals if all of its roots can be found by rational operations and extractions of a root, these operations being performed a finite number of times upon the coefficients of the equation or upon quantities obtained from them by those operations. Every cubic or quartic equation is solvable by radicals in view of Ch. VII. One of our chief aims is to decide whether or not every equation of the fifth or higher degree is solvable by radicals.

The above definition permits the use of the operation of finding one of the pth roots of a quantity previously determined, but not the use of the operations of finding all its pth roots. The use of the latter operations would imply a knowledge of all the pth roots of unity, whereas one of our aims is to prove that, when p is a prime, they are all expressible in terms of radicals whose indices are $< p$.

The solution of an equation solvable by radicals is usually accomplished by the solution of a series of auxiliary equations the roots of any one of which can be found by rational operations and root extractions performed upon its coefficients and the coefficients and roots of the preceding equations of the series. In order to focus our attention upon a particular equation of the series and to have a flexible phraseology, it will prove convenient to employ the following generalization of the foregoing definition of solvability by radicals.

An equation with coefficients in a field $R(k_1, \ldots, k_m)$ shall be said to be *solvable by radicals relatively to that field* if all of its roots can be found by rational operations and extractions of a root performed upon $k_1, \ldots, k_m$ or upon quantities derived from them by those operations.

For example, $x^{13} = 2^{13}$ is evidently solvable relatively to $R(\rho)$, where ρ is a particular imaginary 13th root of unity.

97. Resolvent equations. Let G be the group for a field F of a given equation $f(x) = 0$ with coefficients in F. A rational function ψ of the roots with coefficients in F belongs to a certain subgroup H of index u under G. By Theorem 9 of §91, ψ is a root of a resolvent equation of degree u with coefficients in F. Assume that this equation is solvable by radicals relatively to F. By adjoining its root ψ to F, we obtain a field $F_1 = F(\psi)$ for which the group of $f(x) = 0$ is now H (§92). Assume that successive such adjunctions lead to a field F_k for which the group of $f(x) = 0$ is the identity. Then the roots all belong to F_k by property A of §89.

The assumptions just made will be shown to hold true when $f(x)$ is $x^3 + bx^2 + cx + d$, if b, c, d are independent variables, and F is the field $R(\omega, b, c, d)$, where $\omega = -\frac{1}{2} + \frac{1}{2}(-3)^{\frac{1}{2}}$ is an imaginary cube root of unity. By §93, G is the symmetric group G_6 on the roots x_1, x_2, x_3. To the cyclic subgroup C_3 belongs the alternating function

$$\partial = (x_1 - x_2)(x_2 - x_3)(x_3 - x_1),$$

whose square is the discriminant Δ of the cubic. The resolvent equation $\partial^2 = \Delta$ is solvable by radicals relatively to F. Adjoining $\partial = \Delta^{\frac{1}{2}}$ to F, we obtain a field $F_1 = F(\partial)$ for which the group of our cubic equation is C_3. We employ the linear functions ϕ and ψ of §66 and recall that, in the exercise following it, we verified that the substitution (132) of C_3 replaces ϕ by $\omega\phi$ and ψ by $\omega^2\psi$. Thus all the substitutions of C_3 leave ϕ^3 and ψ^3 unaltered. The latter are equal to rational functions of ∂ with coefficients in F (Theorem 10 of §91). The computations in §§66, 67 give

$$\phi^3 = \tfrac{1}{2}[B + 3\partial(-3)^{\frac{1}{2}}] = g, \quad \psi^3 = \tfrac{1}{2}[B - 3\partial(-3)^{\frac{1}{2}}] = h,$$
$$\phi\psi = b^2 - 3c,$$

where $B = 9bc - 2b^3 - 27d$. The resolvent equation $\phi^3 = g$, where g belongs to the field $F_1 = F(\partial)$, is evidently solvable by radicals relatively to F_1. In the enlarged field $F_2 = F_1(g^{\frac{1}{3}})$, the group of the given cubic equation is the identity group to which ϕ belongs. Hence its roots x_i all belong to F_2; their expressions in terms of quantities in F_2 are given by formulas (5) of §66.

98. Group of a resolvent equation. In our later discussion of the solvability by radicals of the resolvent equation defined at the beginning of §97, we shall need its group. Let ψ belong to the subgroup H of index u under G. In §91 we proved by means of

(1) $$G = H + Hg_2 + Hg_3 + \cdots + Hg_u$$

that ψ is one of exactly u distinct conjugates under the group G:

(2) $$\psi, \psi_{g_2}, \psi_{g_3}, \ldots, \psi_{g_u}$$

and that any substitution s of G replaces these by

(3) $$\psi_s, \psi_{g_2 s}, \psi_{g_3 s}, \ldots, \psi_{g_u s},$$

which are merely the functions (2) rearranged. Hence to any substitution s of G on the letters $x_1, \ldots, x_n$ there corresponds the substitution

(4) $$\sigma = \begin{pmatrix} \psi & \psi_{g_2} & \cdots & \psi_{g_u} \\ \psi_s & \psi_{g_2 s} & \cdots & \psi_{g_u s} \end{pmatrix} \equiv \begin{pmatrix} \psi_{g_i} \\ \psi_{g_i s} \end{pmatrix}$$

on the u letters (2). Similarly, to t of G corresponds

$$\tau = \begin{pmatrix} \psi_{g_i} \\ \psi_{g_i t} \end{pmatrix} \equiv \begin{pmatrix} \psi_{g_i s} \\ \psi_{g_i st} \end{pmatrix},$$

where in the upper row of the final symbol we have adopted the order (3) of the letters (2). This is permissible since a substitution is not changed if we rearrange the columns in its two-rowed notation. The product $\sigma\tau$ therefore replaces ψ_{g_i} by $\psi_{g_i st}$ and hence is equal to a substitution of type (4) which is derived from

(4) by replacing s by st. In other words, the set of all substitutions (4) is a group Γ. This group is transitive since s may be taken to be any g_i.

THEOREM 1. *To each substitution s of G corresponds a unique substitution σ defined by* (4). *All such substitutions σ form a transitive group Γ.*

If to t corresponds τ, we saw that to st corresponds $\sigma\tau$. In other words, the correspondence between substitutions of G and those of Γ is preserved under multiplication. For this reason, G and Γ are said to be *isomorphic*. According as the order of G is equal to or exceeds the order of Γ, G is called *simply* or *multiply* isomorphic to Γ.

EXAMPLE. Let G be the alternating group on the independent variables $x_1, \ldots, x_4$. We have

$$G = G_4 + G_4(234) + G_4(243),$$
$$G_4 = \{I, (12)(34), (13)(24), (14)(23)\}.$$

The indicated substitutions of order 3 replace $\psi = (x_1 - x_2)(x_3 - x_4)$ by $\psi_2 = (x_1 - x_3)(x_4 - x_2)$ and $\psi_3 = (x_1 - x_4)(x_2 - x_3)$. Each substitution of G_4 leaves ψ, ψ_2, and ψ_3 unaltered.[1] Hence ψ belongs to G_4 and has only three conjugates under G. We have

$$\Gamma = \{I, \quad (\psi \psi_2 \psi_3), \quad (\psi \psi_3 \psi_2)\}.$$

For, to each substitution of G_4 corresponds the identity of Γ. To all substitutions of the set $G_4(234)$ therefore corresponds the same substitution $(\psi \psi_2 \psi_3)$ of Γ. Hence G is (4, 1) isomorphic to Γ.

EXERCISES

1. If G is the symmetric group and ψ is the alternating function, $\Gamma = \{I, (\psi\psi_2)\}$.

2. If G is the symmetric group on $x_1, \ldots, x_4$, the group Γ on $y_1 = x_1 x_2 + x_3 x_4$, y_2, y_3 is the symmetric group of order 6.

[1] This was shown in Ex. 4 of §73. See (22) of §71.

3. If G is the symmetric group on x_1, x_2, x_3, and if ψ is the six-valued function $x_1 + \omega^2 x_2 + \omega x_3$, Γ is simply isomorphic to G (cf. §67, Ex. 5).

4. The group Γ is identical with that obtained by using another rational function ψ belonging to H and having its coefficients in F. Hints: By §91, ϕ is a rational function $r(\psi)$ of ψ with coefficients in F. By property B, $\phi - r(\psi)$ is unaltered by every substitution t of G, whence $\phi_t = r(\psi_t)$. Hence to any substitution on the u letters (2) corresponds the same substitution on the distinct letters $\phi, \phi_{\sigma_2}, \ldots, \phi_{\sigma_u}$. The converse is true since we may interchange the rôles of ϕ and ψ.

5. Give another proof that Γ is determined by G and H alone, and is independent of ψ. If s is any substitution of G, prove that the products of the u sets in (1) by s on the right are the same sets rearranged, and that the resulting substitution on the u sets is the same as (4) apart from the notation of the letters operated on.

99. The special importance of the group Γ is due to

THEOREM 2. Γ *is the group for F of the resolvent equation*

$$(5) \qquad g(y) \equiv (y - \psi_1)(y - \psi_2) \cdots (y - \psi_u) = 0$$

with coefficients in F. Here ψ_i denotes ψ_{σ_i}.

To prove that Γ has the characteristic properties A, B of the group of (5) for F, note that any rational function $R(\psi_1, \ldots, \psi_u)$ with coefficients in F is equal to a rational function $r(x_1, \ldots, x_n)$ with coefficients in F:

$$(6) \qquad R(\psi_1, \ldots, \psi_u) - r(x_1, \ldots, x_n) = 0.$$

Since this difference is equal to the number zero of F, it is unaltered by every substitution s of the group G on $x_1, \ldots, x_n$. To s corresponds a definite substitution σ of the group Γ on $\psi_1, \ldots, \psi_u$. Hence

$$(7) \qquad R_\sigma(\psi_1, \ldots, \psi_u) - r_s(x_1, \ldots, x_n) = 0.$$

First, let R be unaltered by every substitution of Γ, so that $R = R_\sigma$ for every σ in Γ. Comparing (6) with (7), we conclude that $r = r_s$ for every s in G. Hence by property A for the group G, r belongs to F. This proves property A for the group Γ.

Second, let R belong to F. By (6), r belongs to F. Hence by property B for the group G, $r_s = r$ for every s in G. Then by (6) and (7), $R_\sigma = R$ for every σ in Γ. This proves property B for the group Γ.

Since Γ is transitive, equation (5) is irreducible in F.

100. Invariant subgroup. Let ψ belong to the subgroup

$$H = \{h_1 = I, h_2, \ldots, h_p\}$$

of G. Let h and s be any substitutions of H and G respectively. Then $t = s^{-1} hs$ leaves ψ_s unaltered, since s^{-1} replaces ψ_s by ψ, h leaves ψ unaltered, and s replaces ψ by ψ_s. Conversely, if t is any substitution of G which leaves ψ_s unaltered, so that $\psi_{st} = \psi_s$, then

$$\psi_{sts^{-1}} = \psi_{ss^{-1}} = \psi_I = \psi, \qquad sts^{-1} = h,$$

where h is in H, whence $t = s^{-1} hs$. Hence ψ_s belongs to the subgroup

$$\{s^{-1} h_1 s = I, \quad s^{-1} h_2 s, \ldots, \quad s^{-1} h_p s\}$$

of G. This subgroup is denoted by $s^{-1} Hs$ and is called the *transform* of H by s. Its substitution $s^{-1} h_i s$ is called the *transform* of h_i by s (cf. §101).

THEOREM 3. *Let ψ belong to the subgroup H of index u under G. By means of (1), we obtain a complete set of conjugates to ψ under G:*

$$\psi, \psi_{\sigma_2}, \ldots, \psi_{\sigma_u}.$$

They belong to the respective groups

$$H, g_2^{-1} H g_2, \ldots, g_u^{-1} H g_u.$$

The latter groups are said to form a complete set of conjugate subgroups of G. In case they are all identical, H is called an *invariant*[1] (or *self-conjugate*) subgroup of G. In this important case, the substitution (4) is the identity if s is in H, since s then leaves each ψ_{g_i} unaltered; while any substitution s of G and the product hs, where h is in H, both correspond to the same substitution (4). In other words, if we consider the sets H, Hg_2, etc., which by (1) make up all the substitutions of G, we see that all substitutions of H correspond to the identity of the group Γ, all of Hg_2 correspond to the same substitution of Γ, etc. To obtain Γ it therefore suffices to take the substitutions which correspond to $I, g_2, \ldots, g_u$. Since these replace ψ by u distinct functions, the resulting u substitutions of Γ are distinct. This proves

THEOREM 4. *If H is an invariant subgroup of G of index u, the group Γ is a transitive group of order u on u letters and hence is a regular group.*

If u is a prime and σ is any substitution $\neq I$ of Γ, the order of σ is u (§76, Exs. 2, 3), so that the powers of σ give all the substitutions of Γ. This proves the

COROLLARY. *If H is an invariant subgroup of G of prime index u, then Γ is a regular cyclic group of order u.*

This corollary is illustrated by Ex. 1 and the example preceding it in §98. For each we shall prove in the next section that H is invariant in G.

---

[1] In other words, H is invariant in G if H is transformed into itself by every substitution of G.

101. Transforms of a substitution. There is a simple rule for finding the transform $g^{-1}hg$ of h by g without performing the indicated multiplications. Let h be a product $rst\cdots$ of cycles affecting different sets of letters. Evidently

$$g^{-1}hg = g^{-1}rg \cdot g^{-1}sg \cdots.$$

It remains to find the transform of a cycle $r = (abc\ldots e)$ by

$$g = \begin{pmatrix} a & b & c & \cdots & e & f & \cdots & l \\ A & B & C & \cdots & E & F & \cdots & L \end{pmatrix}.$$

Then

$$g^{-1} = \begin{pmatrix} A & B & C & \cdots & E & F & \cdots & L \\ a & b & c & \cdots & e & f & \cdots & l \end{pmatrix},$$

$$g^{-1}rg = \begin{pmatrix} A & B & C & \cdots & E & F & \cdots & L \\ B & C & D & \cdots & A & F & \cdots & L \end{pmatrix}.$$

Hence $g^{-1}rg = (ABC\ldots E)$ may be obtained by replacing each letter of the cycle $r = (abc\ldots e)$ by the letter by which g replaces it, or expressed briefly by applying g within the cycle. Hence $g^{-1}hg$ is obtained by applying g within the cycles of h. For example,

$$(123)^{-1} \cdot (12)(34) \cdot (123) = (23)(14).$$

Since any substitution transforms a product of k transpositions into a product of k transpositions, and therefore an even substitution into an even substitution, it follows that the alternating group on n letters is invariant in the symmetric group. The case $k = 2$ shows that

$$G_4 = \{I, (12)(34), (13)(24), (14)(23)\}$$

§ 102] SIMPLE AND QUOTIENT GROUPS 187

is invariant under the symmetric group G_{24} on four letters, since G_4 contains every substitution $(ab)(cd)$, where a, b, c, d form a permutation of 1, 2, 3, 4.

A substitution is called *invariant* in G if it is transformed into itself by every substitution of G.

Exercises

1. G_4 is an invariant subgroup of the alternating group on four letters and also of the groups G_8, H_8, K_8 of §73.

2. If H is a subgroup of G of index 2, H is invariant in G. Hint: $G = H + Hg = H + gH$, whence the substitutions of the set Hg form a permutation of those of gH.

3. The only invariant substitutions of G_8 in §73 are the identity I and $a = (12)(34)$. Hence the only invariant subgroups of order < 4 of G_8 are the identity and $\{I, a\}$.

102. Simple and quotient groups. A group is called *simple* if it has no invariant subgroup other than itself and the identity group. A *composite* group is one which is not simple. A group whose order is a prime is evidently simple.

Let H be an invariant subgroup of index u of a group G. By §91 there exists a function ψ belonging to the subgroup H. Then the group Γ defined in §98 is of order u (Theorem 4 of §100); it is called the *quotient group* of G by H and denoted by G/H. Its order is the quotient u of the order of G by that of H.

For example, if G_6 and G_3 are the symmetric and alternating groups on three letters, the alternating function ψ_1 belongs to G_3 and it takes a second value $\psi_2 = -\psi_1$ under G_6. Then G_6/G_3 is the group $\Gamma = \{I, (\psi_1 \psi_2)\}$.

A *proper* subgroup of G is a subgroup distinct from G.

In particular, let H be a *maximal* invariant proper subgroup of G, so that H is not contained in a larger invariant proper subgroup of G.

THEOREM 5. *The quotient group* $\Gamma = G/H$ *is simple.*

For, suppose that Γ has an invariant subgroup Δ distinct from Γ and the identity group. Write $\Delta = \{\alpha_1 = I, \alpha_2, \ldots, \alpha_d\}$, where α_i corresponds to all the substitutions of a set Ha_i forming a component of G in the notation (1). Write

$$D = Ha_1 + Ha_2 + \cdots + Ha_d, \qquad a_1 = I.$$

Since Δ is a group, $\alpha_i \alpha_j = \alpha_k$; whence $Ha_i \cdot Ha_j = Ha_k$, so that D is a group. To every substitution g of G corresponds a substitution γ of Γ. Since H is invariant in G, all the substitutions of the set $g^{-1} Ha_i g = H \cdot g^{-1} a_i g$ correspond to $\gamma^{-1} \alpha_i \gamma$, which is a substitution α_j of the invariant subgroup Δ of Γ. But the substitutions of Ha_j alone correspond to α_j. Hence $g^{-1} Ha_i g = Ha_j$, so that D is invariant in G. Since Δ is distinct from Γ and the identity, D is distinct from G and H. Hence H is contained in the larger invariant proper subgroup D of G, contrary to hypothesis.

103. Series and factors of composition, solvable groups. Any group G contains the identity group G_1, which is invariant in G. Since there is only a finite number of sets of substitutions in a group G of finite order, G has a finite number of subgroups. These two remarks show that G has a maximal invariant proper subgroup H. Similarly, H has a maximal invariant proper subgroup K. Such a series of groups $G, H, K, \ldots, M, G_1$, terminating with G_1, is called a *series of composition* of G. If H is of index u under G, K of index v under H, $\ldots$, and if m is the index of G_1 under M (i.e., the order of M), the positive integers $u, v, \ldots, m$ are called the *factors of composition* of G (relative[1] to the series $G, H, \ldots, G_1$).

---

[1] But in fact the same for all series of composition of G, a result due to Jordan, Traité des Substitutions, 1870, 42. It is a corollary to the theorem of Hölder (Math. Annalen, 34, 1889, 37; simplified by G. L. Brown, Bull. Amer. Math. Soc., 1, 1894–5, 232–4) that the quotient groups $G/H, H/K, \ldots, M$ are identical in some order with those of any series of composition of G. Cf. Miller, Blichfeldt, and Dickson, Theory and Applications of Finite Groups, 1916, 175–6.

§ 103]   SERIES AND FACTORS OF COMPOSITION   189

A group is called *solvable* if its factors of composition are all primes, otherwise *insolvable*.

EXERCISES

1. If G_6 and G_3 are the symmetric and alternating groups on 3 letters, the only series of composition of G_6 is G_6, G_3, G_1, and G_6 is a solvable group.

2. A cyclic group of prime order is a solvable group.

3. The symmetric group on 4 letters has the series of composition G_{24}, G_{12} = alternating group, G_4 (in Ex., §98), G_2, G_1, there being three choices for G_2. Hence G_{24} is a solvable group.

4. If C_{12} is generated by $a = (x_1 x_2 \cdots x_{12})$, the only proper subgroups are $C_6 = \{I, a^2, a^4, a^6, a^8, a^{10}\}$, $C_4 = \{I, a^3, a^6, a^9\}$, $C_3 = \{I, a^4, a^8\}$, $C_2 = \{I, a^6\}$, and $C_1 = \{I\}$. Every subgroup of a commutative group is invariant. The only series of composition of C_{12} are

$$C_{12}, C_6, C_3, C_1; \quad C_{12}, C_6, C_2, C_1; \quad C_{12}, C_4, C_2, C_1.$$

The factors of composition are 2, 2, 3; 2, 3, 2; 3, 2, 2, respectively.

5. If G has a subgroup H of index 2, H a subgroup K of index 2, etc., to the identity group G_1, then G has the series of composition $G, H, K, \ldots, G_1$.

6. Find the seven series of composition of G_8 in §73.

7. Prove that the three series of composition in Ex. 3 are the only ones.

104. Theorem 6. *The solution of an equation having the group G for the field F can be reduced to the solution of a series of equations each having a simple regular group for the field obtained by adjoining to F a root of each of the earlier equations of the series. In particular, if G is a solvable group, each auxiliary equation has a regular cyclic group of prime order.*

Let $G, H, K, \ldots, G_1$ be a series of composition of G, and $u, v, \ldots$, be the corresponding factors of composition. Construct a rational function ψ of the roots x_i with coefficients in F such that ψ belongs to the subgroup H of G of index u (§91). Then ψ is a root of an equation of degree u with coefficients in F whose group Γ for F is simply isomorphic with the simple quotient group

G7H (§§99, 102). After the adjunction of the root ψ to F, the group of the given equation becomes H for the field $F(\psi)$ by §92.

Construct a rational function χ of the x_i with coefficients in $F(\psi)$ such that χ belongs to the subgroup K of H of index v. Then χ is a root of an equation of degree v with coefficients in $F(\psi)$ whose group for that field is simply isomorphic with the simple group H/K. By the adjunction of χ, the group of the given equation becomes K for the field $F(\psi, \chi)$.

The final such step is the adjunction of a function belonging to G_1. The resulting field contains each x_i.

105. Equations with a cyclic group. The last result is supplemented by the following

THEOREM 7. *An equation, whose group G for a field F containing an imaginary pth root ρ of unity is a regular cyclic group of prime order p, is solvable by radicals relatively to F.*

Let the roots be $x_0, x_1, \ldots, x_{p-1}$ and let G be generated by the substitution $s = (x_0 \, x_1 \, \cdots \, x_{p-1})$. We see that s replaces

$$(8) \qquad r_i = x_0 + \rho^i x_1 + \rho^{2i} x_2 + \cdots + \rho^{(p-1)i} x_{p-1}$$

by $\rho^{-i} r_i$ and hence leaves $v_i = r_i^p$ unaltered. Since the coefficients of v_i belong to F and it is unaltered by all the substitutions s^k of G, v_i is equal to a quantity in F by property A of §89. Evidently $-r_0$ is a coefficient of the given equation. The p linear equations (8) for $i = 0, 1, \ldots, p-1$ can be solved for the p unknowns $x_0, \ldots, x_{p-1}$ by the method employed in §66 for the case $p = 3$. Multiply (8) by ρ^{-ij}, where $0 \leq j < p$, and sum for $i = 0, 1, \ldots, p-1$. In the resulting relation, the coefficient of x_j is $1 + \cdots + 1 = p$, while, for $k \neq j$, the coefficient of x_k is

$$\sum_{i=0}^{p-1} \rho^{ti} = \frac{\rho^{tp} - 1}{\rho^t - 1} = 0 \qquad (t = k - j),$$

since l is not divisible by p. Hence[1]

$$px_j = r_0 + \sum_{i=1}^{p-1} \rho^{-ij} v_i^{1/p} \qquad (j = 0, 1, \ldots, p-1).$$

106. Cyclotomic equations. To perfect the last theorem, it remains to prove that we can find by root extractions the imaginary pth roots of unity, where p is an odd prime. They are the successive powers of one of them, say ρ. For $p = 5$, they may be arranged in the order

$$\rho, \ \rho^2, \ \rho^{2^2} = \rho^4, \ \rho^{2^3} = \rho^3,$$

so that each is the square of its predecessor.

Since the discussion does not become more complicated, we shall treat the generalization from p to p^s and hence obtain results needed also in our later study of regular polygons. In §83 we saw that the primitive ninth roots of unity are

$$\rho, \ \rho^2, \ \rho^4, \ \rho^8, \ \rho^{16} = \rho^7, \ \rho^{32} = \rho^5,$$

each of which is the square of its predecessor, and that the $e = p^s - p^{s-1}$ primitive p^sth roots of unity are those powers of one of them, say ρ, whose exponents are the positive integers less than p^s and not divisible by p. The above facts concerning fifth and ninth roots of unity may be generalized as follows: It is shown in the theory of numbers that there exists an integer g such that, when $1, g, g^2, \ldots, g^{e-1}$ are divided by p^s, the positive remainders $< p^s$ coincide in some order with all the positive integers $< p^s$ which are not divisible by p, and hence give the mentioned exponents of ρ. In other words, the primitive p^s roots of unity are

$$x_1 = \rho, \ x_2 = \rho^g, \ x_3 = \rho^{g^2}, \ldots, x_e = \rho^{g^{e-1}}.$$

[1] Lagrange, Traité résolution équations, 1808, Note XIII, §16. Implicitly in Nouv. Mém. Acad. Berlin, année 1770, §71; Œuvres, III, 333.

Hence

(9) $\quad x_2 = x_1{}^g,\ x_3 = x_2{}^g,\ \ldots,\ x_e = x^g{}_{e-1},\ x_1 = x_e{}^g.$

The final relation follows from Euler's theorem that g^e is of the form $1 + kp^s$, where k is an integer (as shown in the theory of numbers). Compare the above cases $p^s = 5$, $p^s = 9$.

We proved in §83 that the equation having these roots $x_1, \ldots, x_e$ is irreducible in the field R of rational numbers. Hence its group G for R is transitive. Let

$$s = \begin{pmatrix} x_1 & x_2 & x_3 \cdots x_e \\ x_a & x_b & x_c \cdots x_l \end{pmatrix}$$

be any substitution of G. By property B of §89, it follows from (9) that

$$x_b = x_a{}^g,\ x_c = x_b{}^g,\ \ldots,\ x_a = x_l{}^g.$$

But $x_{a+1} = x_a{}^g$, etc., by (9). Hence

$$x_b = x_{a+1},\ x_c = x_{b+1},\ \ldots,\ x_a = x_{l+1},$$

provided the new symbol x_{e+1} be replaced by x_1. If two integers u and v differ by a multiple of e, we write $u \equiv v \pmod{e}$. Hence

$$b \equiv a + 1,\ c \equiv b + 1 \equiv a + 2, \ldots \pmod{e},$$

$$s = \begin{pmatrix} x_1 & x_2 & x_3 & \cdots x_k & \cdots x_e \\ x_a & x_{a+1} & x_{a+2} & \cdots x_{k+a-1} & \cdots x_{e+a-1} \end{pmatrix},$$

in which x_{e+j} is to be replaced by x_j. Hence

$$s = t^{a-1},\ t = (x_1\ x_2\ x_3\ \cdots\ x_e).$$

Since G is transitive it contains all the powers of t.

Theorem 8. *If p is an odd prime, the group for the field of rational numbers of the cyclotomic equation whose roots are the primitive p^sth roots of unity is a regular cyclic group of order $e = p^s - p^{s-1}$.*

This theorem was verified for $p^s = 5$ in Ex. 8, §90.

107. Sufficient condition for solvability by radicals. An equation having a solvable group for the field determined by the coefficients is solvable by radicals. We shall prove the more general

THEOREM 9. *An algebraic equation having a solvable group for any field F containing the coefficients is solvable by radicals relatively to F.*

Assume for the moment that the theorem holds for every equation having for F a regular cyclic group of prime order. We shall then prove that any equation $f(x) = 0$ whose group for F has prime factors of composition $u, v, \ldots$ is solvable by radicals relatively to F. As in §104, there is a series of equations $n(\psi) = 0$, $m(\chi) = 0, \ldots$ of prime degrees $u, v, \ldots$, the solution of which is equivalent to the solution of $f(x) = 0$. As there proved, the group for F of $n(\psi) = 0$ is a regular cyclic group of prime order u, which by our assumption is solvable by radicals relatively to F. The coefficients of $m(\chi) = 0$ belong to the field $F_1 = F(\psi)$ and its group for F_1 is a regular cyclic group of prime order v (§104); by our assumption it is solvable by radicals relatively to F_1. It is therefore solvable by radicals relatively to F, since the same was just proved for $n(\psi) = 0$. Similarly, all of our auxiliary equations are solvable by radicals relatively to F. Since the roots of $f(x) = 0$ are expressible rationally in terms of quantities in F and roots $\psi, \chi, \ldots$ of these auxiliary equations (§104), we see that $f(x) = 0$ is solvable by radicals relatively to F.

It remains to justify the assumption made above that any equation $C(x) = 0$ having a regular cyclic group G of prime order p for a field F is solvable by radicals of indices $\leq p$ relatively to F. This is evidently true when $p = 2$. To proceed by induction on p, suppose that every equation having a regular cyclic group of order q, where q is a prime $< p$, for any field D is solvable by radicals of indices $\leq q$ relatively to D. As in the preceding paragraph, this implies that the equation for the imaginary pth roots of unity is solvable by radicals of indices $< p$ (i.e., relatively to the field R of rational numbers). In fact, its group for R is a regular cyclic group of order $p - 1$ (§106), whose factors of composition are the prime factors of $p - 1$, and these primes are $< p$.

Adjoin to F an imaginary pth root ρ of unity. The group of $C(x) = 0$ for the field $F(\rho)$ is either the initial cyclic group G of prime order p or the identity group (§92). In the latter case, the roots are in $F(\rho)$ and can be found from quantities in F by rational operations and extractions of roots of indices $< p$, since ρ was seen to be derivable from rational numbers by those operations, so that $C(x) = 0$ is solvable by radicals of indices $< p$ relatively to F. In the former case, $C(x) = 0$ is solvable (§105) by radicals of indices $\leqq p$ relatively to $F(\rho)$ and hence, as before, relatively to F.

Hence the induction is complete and the theorem is proved.

COROLLARY. *If p is an odd prime, the equation for the $p - 1$ imaginary pth roots of unity is solvable by radicals of indices $< p$.*

EXAMPLE 1. If b, c, d are independent variables and ω is an imaginary cube root of unity, the group of $x^3 + bx^2 + cx + d = 0$ for the field $R(\omega, b, c, d)$ is the symmetric group on the three roots (§93). Since it is a solvable group by Ex. 1, §103, the equation is solvable by radicals (cf. §97). Note that after the adjunction of the square root of the discriminant, the equation has a cyclic group of prime order 3.

EXAMPLE 2. Consider the quartic equation whose coefficients are independent variables. For the field F determined by its coefficients, the group is the symmetric group G_{24} on the roots. It is a solvable group by Ex. 3, §103. To the invariant subgroup G_{12}, composed of the even substitutions, belongs the product P of the differences of the roots. The square of P is the discriminant Δ of the quartic equation. The group reduces to G_{12} for the field $F_1 = F(P)$. The function $y_1 = x_1 x_2 + x_3 x_4$ is unaltered by $G_4 = \{I, (12)(34), (13)(24), (14)(23)\}$, while the remaining eight substitutions of G_{12} replace y_1 by y_2 or y_3. These three y's are by §91 the roots of a cubic equation with coefficients in F_1 (the resolvent cubic in §71). Since G_4 is invariant in G_{12}, the group of this cubic for the field $F(P)$ is G_{12}/G_4, which is a cyclic group of order 3. By the adjunction of y_1 to F_1, the group of the quartic reduces to G_4. The function $x_1 x_2$ is unaltered by $G_2 = \{I, (12)(34)\}$, while the remaining substitutions of G_4 replace it by $x_3 x_4$. The adjunction of a root of $z^2 -$

§ 107] EXERCISES 195

$y_1 z + x_1 x_2 x_3 x_4 = 0$ reduces G_4 to the invariant subgroup G_2. Finally, the adjunction of $x_1 - x_2$ whose square belongs to $F_1(y_1, x_1 x_2)$ reduces G_2 to the identity group. The enlarged field contains $x_1, \ldots, x_4$, which have been found by the extraction of three square roots and one cube root.

EXERCISES

1. Instead of y_1 we may employ in Example 2

$$t_1^2 = 4y_1 + b^2 - 4c, \qquad t_1 \equiv x_1 + x_2 - x_3 - x_4,$$

where the notations of §71 are used. Evidently the transformation $y = \frac{1}{4}(t - b^2 + 4c)$ replaces the former resolvent cubic by one in t whose roots are the squares of t_1, $t_2 = x_1 + x_3 - x_2 - x_4$, $t_3 = x_1 + x_4 - x_2 - x_3$. From these three linear relations and $-b = \sum x_1$, determine the roots x_i.

2. Modify Ex. 2 by adjoining the cube root ω of unity to the initial field and employing instead of y_1 the function $\phi = y_1 + \omega y_2 + \omega^2 y_3$. Write $\psi = y_1 + \omega^2 y_2 + \omega y_3$. Applying to the resolvent cubic (19) in §71 the results in §66 with x's changed to y's, show that

$$\phi^3 - \psi^3 = 3(\omega - \omega^2)P, \qquad \phi\psi = c^2 - 3bd + 12e,$$
$$\phi^3 + \psi^3 = 2c^3 - 9bcd - 72ce + 27d^2 + 27b^2 e,$$

which are essentially the invariants I and J of §12. Hence ϕ is found as a cube root and then ψ is known. From the values of ϕ, ψ, and $y_1 + y_2 + y_3 = c$, we obtain y_1, y_2, y_3 by solving three linear equations (cf. §66). For the field F_1 obtained by adjoining P and ϕ, the group of the quartic equation is G_4. If also the root t_1 of $t_1^2 = 4y_1 + b^2 - 4c$ is adjoined, the group reduces to $G_2 = \{I, (12)(34)\}$. Finally, if also a root t_2 of $t_2^2 = 4y_2 + b^2 - 4c$ is adjoined, the group reduces to the identity and the x's may be found as in Ex. 1. Note that $t_1 t_2 t_3 = 4bc - b^3 - 8d$. The elegance of this solution is due to its employment of binomial resolvents only and to its introduction of the invariants I and J.

108. Jordan's[1] theorem on the mutual adjunction of roots.

THEOREM 10. *Let the group G_1 for a field F of an algebraic equation $f_1(x) = 0$ be reduced to H_1 by the adjunction of all of the roots of a second equation $f_2(x) = 0$, and let the group G_2 for F of the second equation be reduced to H_2 by the adjunction of all of the*

[1] Traité des Substitutions, 1870, 268–9.

roots of the first equation. Then H_1 and H_2 are invariant subgroups of G_1 and G_2, respectively, of equal indices, and[1] *the quotient groups G_1/H_1 and G_2/H_2 are simply isomorphic.*

By §91, there exists a rational function g_1 with coefficients in F of the roots $x_1, \ldots, x_n$ of the first equation such that g_1 belongs to the subgroup H_1 of G_1. Since the adjunction of the roots $y_1, \ldots, y_m$ of the second equation reduces G_1 to H_1, property A of H_1 (§89) shows that g_1 is a quantity of the enlarged field, whence

$$(10) \qquad g_1(x_1, \ldots, x_n) = h_1(y_1, \ldots, y_m),$$

where h_1 is a rational function with coefficients in F. Let $g_1, \ldots, g_k$ denote the distinct functions obtained from g_1 by applying the substitutions of G_1. Then H_1 is of index k under G_1 (§91). Let $h_1, \ldots, h_l$ denote the distinct functions obtained from h_1 by applying the substitutions of G_2. The k functions g_i are the roots of an equation irreducible in F; likewise for the l functions h_i. Since these two irreducible equations have a root (10) in common, they are identical by §80. Hence $k = l$ and the g's coincide in some order with the h's.

If s_i is a substitution of G_1 which replaces g_1 by g_i, then s_i transforms the group H_1 to which g_1 belongs into the group to which g_i belongs (§100), whose order is that of H_1. Since g_i is equal to a certain h_j it belongs to the field $F' = F(y_1, \ldots, y_m)$ and hence is unaltered by the substitutions of the group H_1 of $f_1(x) = 0$ for that field F' (property B of §89). Hence the group to which g_i belongs contains all the substitutions of H_1 and, being of the same order as H_1, coincides with H_1. This proves that H_1 *is invariant in* G_1. The group for F of the irreducible equation satisfied by g_1 is therefore the quotient group G_1/H_1 (§102).

---

[1] This supplement and the proof here employed are due to Hölder, Math. Annalen, 34, 1889, 47. For generalizations and related results, see Kneser, *ibid.*, 30, 1887, 179–202; shorter proof of theorem on p. 195 in Jour. für Math., 106, 1890, 51. Landsberg, *ibid.*, 132, 1907, 1–20. Bucca, Rendiconti Circolo Mat. Palermo, 14, 1900, 122–6. Loewy, Math. Zeitschrift, 15, 1922, 261–73.

Let K be the subgroup of G_2 to which $h_1(y_1, \ldots y_m)$ belongs. It is of index k since h_1 is a root of an equation of degree $l = k$ irreducible in F. By the adjunction of h_1 to F and hence of g_1 by (10), the group G_2 of $f_2(x) = 0$ for F is reduced to K (§92). If not merely $g_1(x_1, \ldots, x_n)$, but all of the x_i themselves be adjoined, the group G_2 reduces to a subgroup of K. Hence H_2 is contained in K. This proves the preliminary result: If the group of $f_1(x) = 0$ reduces to a subgroup of index k on adjoining all of the roots of $f_2(x) = 0$, then the group of $f_2(x) = 0$ reduces to a subgroup of index k_1 ($k_1 \geqq k$) on adjoining all of the roots of $f_1(x) = 0$.

Interchanging f_1 and f_2 in the preceding statement, we obtain the following result: If the group of $f_2(x) = 0$ reduces to a subgroup of index k_1 on adjoining all the roots of $f_1(x) = 0$, then the group of $f_1(x) = 0$ reduces to a subgroup of index k_2 ($k_2 \geqq k_1$) on adjoining all of the roots of $f_2(x) = 0$. Since the hypothesis for the second result is identical with the conclusion for the first result, we conclude that $k_2 = k$, and $k_1 \geqq k \geqq k_1$, whence $k_1 = k$. Since H_2 is contained in K, and H_2 is of index k_1 under G_2 and K is of index k under G_2, it follows that H_2 and K are identical.

For the same reason that H_1 is invariant in G_1, it follows that H_2 is invariant in G_2. Since h_1 belongs to the subgroup $K = H_2$ of G_2, the group for F of the irreducible equation satisfied by h_1 is the quotient group G_2/H_2. Since this equation was seen to be identical with that satisfied by g_1 and since the group of the latter equation for F was seen to be G_1/H_1, the groups G_2/H_2 and G_1/H_1 differ only in the notations employed for the letters on which they operate, and hence are simply isomorphic.

109. Galois's theorem on adjunction.

THEOREM 11. *By the adjunction of any one root of an equation $f_2(x) = 0$ whose group for a field F is a regular cyclic group of prime order p, the group for F of the equation $f_1(x) = 0$ either is not reduced at all or else is reduced to an invariant subgroup of index p.*

This is a corollary to Jordan's theorem, since if we adjoin one root x_1 of $f_2(x) = 0$ we really adjoin all of its roots. For, the identity is the only substitution of the group for F of $f_2(x) = 0$ which leaves x_1 unaltered in value, whence each root is a rational function of x_1 with coefficients in F (Theorem 10 of §91).

110. Galois's criterion for solvability by radicals. *An algebraic equation is solvable by radicals if and only if its group for the field determined by the coefficients is a solvable group.*

It is occasionally useful to have the generalization:

THEOREM 12. *An equation is solvable by radicals relatively to any field F containing the coefficients if and only if its group for F is a solvable group.*

That the equation is solvable when the group is a solvable group was proved in §107. We shall now prove the converse. Assume therefore that the roots $x_1, \ldots, x_n$ can be derived by rational operations and root extractions from quantities in F or from quantities obtained from them by those operations. The index of each root extraction may be assumed to be a prime since a pqth root is a pth root of a qth root. Such a two-story radical as well as the underneath radical are both listed in making a list of all the radicals

(11) $$a_1^{1/p_1}, \quad a_2^{1/p_2}, \ldots, a_k^{1/p_k}$$

which occur in the expressions for $x_1, \ldots, x_n$. Here a_1 is in F, a_2 is in the field derived from F by adjoining the first radical, a_3 is in the field derived by adjoining the first two radicals, etc. Also $p_1, \ldots, p_k$ are primes.

By the corollary in §107, the equation for the imaginary pth roots of unity (where p is a prime) is solvable by radicals of indices $< p$. Those roots may therefore be expressed rationally in terms of radicals forming a chain of type (11), where now F is the field of

§ 110] GALOIS'S CRITERION FOR SOLVABILITY 199

rational numbers. Write down the radicals of the chain for $p = 3$ and follow them by the radicals of the chain for $p = 5$, and so on for the primes in order up to and including the maximum of $p_1, \ldots, p_k$. After the last of these radicals write the radicals (11) in order. We obtain a chain of radicals

(12) $$b_1^{1/q_1},\ b_2^{1/q_2}, \ldots, b_s^{1/q_s}$$

having the following properties: First, $q_1, \ldots, q_s$ are primes. Second, b_1 is in the field F, while $b_2, b_3, \ldots$ are in the fields derived from F by adjoining the first radical, the first two radicals, ..., respectively. Third, $x_1, \ldots, x_n$ are rational functions with coefficients in F of these radicals, since they were such functions of the included radicals (11). Fourth, after the adjunction to F of the first $r - 1$ radicals (12), we obtain a field containing all the imaginary q_rth roots of unity.

LEMMA. *For a field containing A and an imaginary pth root ρ of unity, where p is an odd prime, the group of $x^p = A$ is the identity group if one root belongs to the field, but is a regular cyclic group of order p if no root belongs to the field.*

For, if one root is r, the remaining roots are $\rho r,\ \rho^2 r, \ldots, \rho^{p-1} r$. All of the latter belong to the field containing ρ if r does, and the group is then the identity. Next, let no root belong to the field, so that A is not the pth power of a quantity of the field. Then $x^p - A$ is irreducible (§82), so that the group is transitive and hence of order $\geq p$. The notation for the roots x_i may be chosen so that

$$x_2 = \rho x_1,\ x_3 = \rho x_2, \ldots, x_p = \rho x_{p-1},\ x_1 = \rho x_p.$$

Consider any substitution

$$s = \begin{pmatrix} x_1 & x_2 & x_3 & \cdots & x_p \\ x_a & x_b & x_c & \cdots & x_l \end{pmatrix}$$

of the group. By property B of §89, we have

$$x_b = \rho x_a = x_{a+1}, \quad x_c = \rho x_b = x_{b+1}, \ldots$$

As in §106, s is the power $a - 1$ of $(x_1 x_2 x_3 \cdots x_p)$, which therefore generates the group.

Let G be the group for F of the given equation having the roots $x_1, \ldots, x_n$. By the Lemma and Galois's theorem (§109), the adjunction of the first radical (12) to F either does not affect G or reduces it to an invariant subgroup of index q_1. The subsequent adjunction of the second radical (12) either does not affect the resulting group or reduces it to an invariant subgroup of index q_2. After the adjunction *seriatim* of all the radicals (12) we have a field containing $x_1, \ldots, x_n$, for which the group is now the identity G_1. Since we passed from G to G_1 through a series of groups each an invariant subgroup of its predecessor of prime index, we have a series of composition of G such that each factor of composition is a prime. Hence G is a solvable group.

111. General equation of degree n > 4 not solvable by radicals. By §93, the group of the general equation of degree n (i.e., one whose coefficients are independent complex variables) is the symmetric group G for the field determined by its coefficients and any constants finite in number. By §101, G has as an invariant subgroup the alternating group, which is composed of all the even substitutions on n letters.

THEOREM 13. *If $n > 4$, the alternating group A on n letters is simple.*

For, suppose A has an invariant subgroup H distinct from the identity group G_1. Of the substitutions of H different from the identity I, consider those which affect the least number of letters. Any such substitution h involves the same number of letters in its various cycles. For, if two cycles involve a and b letters, where $a < b$, h^a is not I and affects fewer letters than h does, since it leaves the a letters unaltered.

Having h, the invariant subgroup H of A contains the transform of h by any even substitution e and hence contains $p = e^{-1} h e \cdot h^{-1}$.

(i) If h contains more than three letters in any cycle, say $h = (1234\ldots)\ldots$, the choice $e = (234)$ gives $p = (124)$, which affects fewer letters than h does.

(ii) If h contains at least two cycles of three letters, say $h = (123)(456)\ldots$, the choice $e = (125)$ gives $p = (15243)$, which affects fewer letters than h does.

(iii) If h contains at least two cycles of two letters, say $h = (12)(34)\ldots$, the choice $e = (123)$ gives $p = (13)(24)$, which affects fewer letters than h does unless $h = (12)(34)$. For the latter h, the choice $e = (125)$ gives $p = (125)$, which affects fewer letters than h does.

Hence the substitutions $\neq I$ of H which affect the least number of letters are all of type (abc), since (ab) is odd and hence is not in H.

Next, any cycle of three letters can be transformed into any other such cycle by an even substitution, so that H contains every cycle of three letters. For, (123) is transformed into (abc) by both of the substitutions

$$s = \begin{pmatrix} 1 & 2 & 3 & 4 & 5 & \cdots & n \\ a & b & c & d & e & \cdots & l \end{pmatrix}, \quad s(de) = \begin{pmatrix} 1 & 2 & 3 & 4 & 5 & \cdots & n \\ a & b & c & e & d & \cdots & l \end{pmatrix},$$

where the dots denote the same letters in the two substitutions, one of which is even and the other odd.

Finally, every even substitution e is expressible as a product of cycles of three letters. For, $e = t_1 t_2 \cdots t_{2k}$, where each t_i is a transposition. If t_1 and t_2 are identical, they cancel. If they have just one letter a in common, $t_1 t_2 = (ab)(ac) = (abc)$. If they have no letter in common,

$$t_1 t_2 = (ab)(cd) = (abc)(cad).$$

Similarly for $t_3 t_4, \ldots, t_{2k-1} t_{2k}$. Hence $H = A$.

If $n > 4$, the symmetric group G on n letters therefore has the series of composition G, A, G_1, with the factors of composition 2 and $\frac{1}{2} \cdot n!$. Since the latter is not a prime, G is not a solvable group. This proves

THEOREM 14. *If $n > 4$, the general equation of degree n is not solvable by radicals. Moreover, its roots cannot be found by rational operations and root extractions performed upon the coefficients and any constants, finite in number, or upon quantities derived from them by those operations.*

EXERCISES

1. If $n > 4$, there is no series of composition of the symmetric group G on n letters other than G, A, G_1. For, if G has an invariant subgroup H other than G_1, the proof of Theorem 13 shows that either $H = A$ or else H contains a transposition (ab). In the latter case H contains every transposition and coincides with G.

2. If the general quintic equation is not solvable by radicals, the same is true of the general equation of degree $n > 5$. Hint: Equate to zero the coefficients of x^j for $j < n - 5$.

3. Hence give another proof of Theorem 14 by amplifying the following new proof of the simplicity of the alternating group G on 5 letters. Let s replace 1, 2, 3, 4, 5 by any assigned permutation a, b, c, d, e of them. If an invariant subgroup H of G contains (123), it contains its transform (abc) by an even s. With $(12)(34)$, H contains its transform $(ab)(cd)$ by s or $(12)s$, one of which is even. With $t = (12345)$, H contains its transform $k = (abcde)$ by s if s is even. But if s is odd, $l = (3245)$ transforms $t^2 = (13524)$ into t, so that ls is even and transforms t^2 into k. Hence H is of order $h = 1 + 20x + 15y + 24z$, where x, y, z take only the values 0 or 1. Also, h is a divisor < 60 of 60. If h has the factor 5, then $z = 1$ and $h = 25$, which is not a divisor of 60. Hence h is a divisor of $\frac{1}{2} 60$ and hence $h = 1$.

4. If for a field F the group of an equation with the roots $x_1, \ldots, x_n$ is generated by $s = (x_1 x_2 \cdots x_n)$, the equation is irreducible and $x_2 = r(x_1)$, where r is a rational function with coefficients in F, and also $x_3 = r(x_2), \ldots,$ $x_n = r(x_{n-1})$, $x_1 = r(x_n)$. Conversely, an irreducible equation in F whose roots satisfy these n relations has a regular cyclic group for F.

5. By supplementing Ex. 3, show that every substitution of the alternating group on 5 letters is conjugate to I, (123), $(12)(34)$, $t = (12345)$, or t^2.

112. Solvable quintics. The Jacobi-Cayley resolvent sextic of a quintic has been found recently by the author[1] by direct, elementary methods. The quintic is solvable by radicals if and only if the sextic has a rational root. He easily deduced the covariant resolvents of Perrin and McClintock.

Berwick[2] computed a resolvent sextic whose coefficients are invariants of the corresponding quintic form.

[1] Bull. Amer. Math. Soc., 31, 1925, 515–23. For this and other resolvents, see Brioschi, Math. Annalen, 13, 1878, 109–60.

[2] Proc. London Math. Soc., (2), 14, 1915, 301–7. Result quoted in Mathews, Algebraic Equations, 1915, 55.

Chapter XI

CONSTRUCTIONS WITH RULER AND COMPASSES

113. The problems to be solved. We shall prove that it is not possible, by the methods of Euclidean geometry, to trisect all angles, or to construct a regular polygon of 7 or 9 sides. For these problems the ancient Greeks sought in vain for constructions with ruler and compasses. We shall specify all the values of n for which a regular polygon of n sides can be so constructed. One such value is 17. The fact that a regular polygon of 17 sides can be constructed with ruler and compasses was not suspected during the twenty centuries from Euclid to Gauss.

114. Analytic criterion for constructibility with ruler and compasses. The first step in our consideration of a problem proposed for construction consists in formulating the problem analytically. In the ancient Delian problem of the duplication of a cube, we take as the unit of length a side of the given cube, and seek the length x of a side of another cube whose volume is double that of the given cube, whence $x^3 = 2$.

For the problem to trisect a given angle A, we employ the trigonometric identity

$$(1) \qquad \cos A = 4 \cos^3 \frac{A}{3} - 3 \cos \frac{A}{3}.$$

Multiply each term by 2 and write x for $2 \cos \frac{1}{3}A$. Thus

$$(2) \qquad x^3 - 3x - 2\cos A = 0.$$

Given A we can construct (with ruler and compasses) a line whose length is the positive value of $\pm \cos A$; it is the adjacent leg of a right triangle, with hypotenuse of unit length, formed by dropping a perpendicular from a point in one side of A to the other side, produced if necessary. If it were possible to trisect angle A, i.e., construct angle $\tfrac{1}{3}A$, we could as before construct a line whose length is $\pm \cos \tfrac{1}{3}A$, and hence a line whose length is double that of the former, viz., $\pm x$.

In each of these problems we have been led to the question of the possibility of constructing a line whose length is $\pm x$, where x is a real root of a cubic equation with known coefficients.

CRITERION.[1] *If x is a real root of a given equation, it is possible to construct with ruler and compasses a line of length $\pm x$ if and only if x can be derived by rational operations and extractions of real square roots performed a finite number of times upon the coefficients or upon numbers obtained from them by those operations.*

First, if x can be derived in the manner indicated in the criterion, we can construct a line of length $\pm x$. For, a rational function of given numbers is obtained from them by additions, subtractions, multiplications, and divisions. The construction of the sum or difference of two segments of straight lines is obvious. To construct segments whose lengths are the product $p = ab$ and quotient $q = a/b$ of the lengths a and b of two given segments, use similar triangles two of whose pairs of corresponding sides are $1, b$ and a, p, and $1, q$ and b, a, respectively. Finally, a segment s of length $n^{\frac{1}{2}}$, where n is positive, may be constructed by drawing a semicircle on a diameter composed of two segments of lengths 1 and n, and drawing a perpendicular s to the diameter at the point which separates the two segments.

Second, suppose that we can construct a line segment of length $\pm x$, when we are given segments representing the coefficients of the equation for x, whose leading coefficient is unity. We may choose any line as the first constructed line OX and choose any point O on it as the point from which we draw another line or about which as center we draw a circle. Take O as the origin and

---

[1] Descartes, La géométrie, Leyde, 1638, livre premier (German transl. by Schlesinger, Berlin, 1894; English transl., Open Court Publishing Company).

OX as the x-axis of a system of rectangular coordinates. The further lines and circles constructed are determined by points which, with the exception of O, are located as the intersections of straight lines and circles. The coordinates of the intersection of two intersecting straight lines are evidently rational functions of the coefficients of their equations. If the straight line $y = mx + b$ intersects the circle

$$(x - p)^2 + (y - q)^2 = r^2,$$

the coordinates of the points of intersection are found by eliminating y, solving the resulting quadratic equation for x, and inserting the roots x into $y = mx + b$; hence the coordinates are obtained from m, b, p, q, r by rational operations and the extraction of a real square root. Finally, two intersecting circles cross at the intersections of one of them with their common chord, so that this case reduces to the preceding. Since each such circle or line other than OX was determined by two points which (with the exception of O) were found as intersections of earlier lines and circles, the coefficients of their equations are obtained by rational operations and extractions of real square roots performed either upon the coefficients of the given equation or upon numbers obtained from them by those operations.

115. Trisection of an angle. To prove that it is not possible to trisect angle $120°$ with ruler and compasses, note that $\cos 120° = -\frac{1}{2}$, so that (2) becomes

(3) $$x^3 - 3x + 1 = 0.$$

By Ex. 3 of §81, this equation is irreducible in the field R of rational numbers. Since its discriminant is 81, the alternating function has a rational value ± 9. Hence the group for R is the alternating group G_3 of order 3. By the adjunction of a square root, the group is either not changed or else is reduced to a sub-

§ 115] TRISECTION OF AN ANGLE 207

group of index 2 (§109). The second alternative is excluded. If a root of the cubic equation could be found by rational operations and extractions of square roots performed upon its coefficients or upon numbers obtained from them by those operations, the adjunction of that root to R would not reduce the group G_3, whereas the adjunction of any root reduces G_3 to the identity group. The above criterion therefore proves that it is not possible to construct a line whose length is the root $2 \cos 40°$ of the cubic equation, i.e., to construct angle $40°$. Hence it is impossible to trisect angle $120°$ with ruler and compasses.[1]

EXERCISES

1. A regular polygon of 9 sides cannot be constructed with ruler and compasses.

2. The duplication of a cube is impossible with ruler and compasses. Hints: The group for R of $x^3 = 2$ is the symmetric G_6. The adjunction of one root reduces it to a group of index 3.

3. It is impossible to trisect an angle whose cosine is $\frac{1}{2}$, $\frac{1}{3}$, $\frac{1}{4}$, $\frac{1}{5}$, or p/q if p and q ($q > 1$) are integers without a common factor and q is not divisible by a cube.

4. It is possible to trisect an angle whose cosine is $(4a^3 - 3ab^2)/b^3$, if the integer a is numerically less than the integer b. Hint: Take $\cos \frac{1}{3}A = a/b$ in (1).

5. It is impossible to construct lines representing the lengths of the edges of a rectangular parallelopiped having a diagonal of length 5, surface area 24, and volume 1, 2, 3, or 5.

6. It is impossible to construct a straight line representing the distance x from the circular base of a hemisphere to the parallel plane which bisects the hemisphere (problem of Archimedes). Hint: The equation for x is (3).

116. Regular polygons. The construction of a regular n-gon by ruler and compasses is equivalent to that of angle $2\pi/n$ and hence of a line of length $\pm \cos 2\pi/n$. The irreducible equation satisfied

---

[1] The author gave an elementary proof without group theory in Amer. Math. Monthly, 21, 1914, 259–62; Mathematics Teacher, 14, 1921, 217–23.

by the latter number is much more difficult to form and treat than that having the root

(4) $$r = \cos 2\pi/n + i \sin 2\pi/n,$$

which is an nth root of unity. We have

$$1/r = \cos 2\pi/n - i \sin 2\pi/n, \quad r + 1/r = 2 \cos 2\pi/n.$$

Hence if r can be expressed in terms of i and real square roots, $\cos 2\pi/n$ can be expressed in terms of real square roots. The converse follows from (4) and the fact that the sine can be found from the cosine by a real square root. Hence a regular n-gon can be constructed with ruler and compasses if and only if the imaginary nth root (4) of unity can be found by extraction of square roots, all except the last one of which is real.

First, let $n = p^s$, where p is an odd prime. Then r is a root of an equation of degree $e = p^{s-1}(p-1)$ whose group for the domain R of rational numbers is a regular cyclic group G of order e (§106). Since the identity is the only substitution of G which leaves a root unaltered, the adjunction of that root reduces G to the identity group G_1. If a regular p^s-gon can be constructed with ruler and compasses, the adjunction of the root r is equivalent to that of several square roots, the adjunction of each of which causes either no reduction in G or a reduction to a subgroup of index 2 (§109). Since the adjunction of r reduces G to G_1, its order e is a power of 2. If $s > 1$, e has the odd factor p and is not a power of 2. If $s = 1$, then $e = p - 1$; if also $e = 2^{at}$, where a is odd, then $p = 2^{at} + 1$ has the factor $2^t + 1$ and is not a prime. Hence *if a regular p^s-gon can be constructed, where p is an odd prime, then $s = 1$ and p is of the form*

(5) $$2^{2^k} + 1.$$

For $k = 0, 1, 2, 3, 4$, the corresponding numbers (5) are 3, 5, 17, 257, 65537, all of which are primes. But for $k = 5, 6, 7, 8, 9, 11, 12$, (5) is known to be not a prime.

§ 116] REGULAR POLYGONS 209

Conversely, if p is a prime such that $p - 1 = 2^h$, a regular p-gon can be constructed with ruler and compasses. For, the group for R of the equation for r in (4) is a regular cyclic group G of order 2^h. Hence G has a series of composition for which each index is 2.

By §104, the solution of the equation for r reduces to the solution of a series of quadratic equations.[1]

Next, let $n = ab$, where a and b are integers without a common factor > 1. If a regular a-gon and a regular b-gon can be constructed with ruler and compasses, the same is true of a regular n-gon. For, any multiples of the angles $2\pi/a$ and $2\pi/b$ can then be constructed and hence also the difference of these multiples. By §78, there exist positive integers c and d such that $ca - db = 1$. Hence we can construct the angle

$$c \cdot \frac{2\pi}{b} - d \cdot \frac{2\pi}{a} = \frac{2\pi}{ab}(ca - db) = \frac{2\pi}{ab}$$

and therefore a regular ab-gon. Conversely, from the latter we obtain a regular a-gon by using the vertices numbered 1, $b + 1$, $2b + 1, \ldots, (a - 1)b + 1$. Hence if $n = p^s q^t \cdots$, where $p, q, \ldots$ are distinct primes, a regular n-gon can be constructed if and only if a regular p^s-gon, a regular q^t-gon, ... can all be constructed. Since a regular 2^l-gon can be constructed by repeated bisections of 180°, our results may be combined into the

THEOREM. *A regular polygon of n sides can be constructed with ruler and compasses if and only if $n = 2^l p_1 p_2 \ldots$, where $p_1, p_2, \ldots$ are distinct primes of the form* (5).

[1] In Dickson's First Course in the Theory of Equations, 1922, 41–44, these quadratic equations are obtained when $p = 17$ and a construction of a regular 17-gon is deduced. The same results are given also in his article in Monographs on Modern Mathematics, Longmans, Green, & Co., 1911, which gives also an exposition of the method of Gauss for finding the series of equations whose solution leads to r for any prime p.

Chapter XII

REDUCTION OF EQUATIONS TO NORMAL FORMS

117. Tschirnhaus transformations. Let $x_1, \ldots, x_n$ be the roots of

(1) $$f(x) \equiv x^n + c_1 x^{n-1} + \cdots + c_n = 0.$$

To find the equation having the roots $y_i = x_i + v$, we have merely to apply the transformation $y = x + v$; we get

$$f(y - v) \equiv y^n + (c_1 - nv)y^{n-1} + \cdots = 0.$$

The coefficient of y^{n-1} is zero if $v = c_1/n$. Hence we can remove the second coefficient by rational operations.

To find the equation whose roots are the squares of those of (1), transpose all terms of odd degrees, square each new member, and then replace x^2 by y.

These transformations $y = x + v$, $y = x^2$, as well as $y = 1/x$ (which is used to find an equation whose roots are the reciprocals of the roots of a given equation), are all cases of a Tschirnhaus transformation

(2) $$y = g(x)/h(x),$$

where g and h are polynomials such that $h(x)$ vanishes for no root of $f(x) = 0$.

Exercises

1. The equation whose roots are the squares of the roots 1, 2, -2 of $x^3 - x^2 - 4x + 4 = 0$ is $y^3 - 9y^2 + 24y - 16 = 0$.

2. Find the equation whose roots are the cubes of those of $x^5 + ax + b = 0$.
Hint:
$$(-x^5)^3 = a^3x^3 + b^3 + 3abx(ax+b).$$
Replace $ax + b$ by $-x^5$ and then write y for x^3.

3. To find the equation whose roots are the kth powers of the roots of (1), employ the kth roots $r_1, \ldots, r_k$ of unity and

$$P = \prod_{i=1}^{k} f(yr_i) = \prod_{i=1}^{k}[(yr_i - x_1) \cdots (yr_i - x_n)]$$

$$= \prod_{j=1}^{n}[(yr_1 - x_j) \cdots (yr_k - x_j)] = (-1)^{(k+1)n} \prod_{j=1}^{n}(y^k - x_j^k).$$

Hence the expansion of the initial product P contains only powers of y whose exponents are multiples of k. Write $y^k = z$.

4. Transformation (2) is only apparently more general than $y = P(x)$, where P is a polynomial. Hint: By §78, there exist polynomials $s(x)$ and $t(x)$ for which $1 \equiv s(x)f(x) + t(x)h(x)$. Take as x a root x_i of $f(x) = 0$.

5. Let $\phi(y) = 0$ be the equation having the roots $y_i = P(x_i)$ for $i = 1, \ldots, n$. If $y_1, \ldots, y_n$ are distinct, then $x_1 = r(y_1), \ldots, x_n = r(y_n)$, where $r(z)$ is a rational function of z whose coefficients are derived rationally from those of $P(x)$ and $f(x)$. Hence if we solve $\phi(y) = 0$ and find that its roots are distinct, we can find the roots of $f = 0$ rationally in terms of the y_i.

Hint: The greatest common divisor $d_i(x)$ of $f(x)$ and $P(x) - y_i$ is found by rational operations, and is of the first degree in x (otherwise it would vanish for two distinct roots x_i and x_j of $f = 0$, and then $P(x) - y_i$ would vanish for $x = x_j$, whence $y_j = y_i$).

6. If $\phi(y) = 0$ in Ex. 5 has a root y_i of multiplicity m_i, the determination of the m_i roots of $f(x) = 0$ for which $P(x)$ has the value y_i requires the solution of an equation $d_i(x) = 0$ of degree m_i.

Hint: For each of the m_i roots, both $f(x)$ and $P(x) - y_i$ vanish, and therefore also $d_i(x)$, which is a linear combination of them (§78).

118. Principal equations. We can always transform (1) into a *principal* equation of degree n in y in which the coefficients of y^{n-1} and y^{n-2} are both zero. By §117, we may assume that $c_1 = 0$. Let $c_2 \neq 0$. Write s_k for the sum of the kth powers of the roots

$x_1, \ldots, x_n$. By Newton's identities, $s_1 = 0$, $s_2 = -2c_2$. Employ the transformation $y = x^2 + ax + b$. Write y_k for the value of y when $x = x_k$. Then $\sum y_k = s_2 + nb$ will be zero if we take $b = -s_2/n$. Next,

$$y^2 = x^4 + 2ax^3 + (a^2 + 2b)x^2 + 2abx + b^2,$$

whence

$$\sum y^2{}_k = s_4 + 2as_3 + (a^2 + 2b)s_2 + nb^2 = 0$$

is a quadratic equation for a in which the coefficient s_2 of a^2 is not zero, and hence has two roots. Choosing either root, we may compute similarly $\sum y^3{}_k$, $\sum y^4{}_k$, ... and then compute by Newton's identities the coefficients of the equation having the roots $y_1, \ldots, y_n$.

THEOREM[1] 1. *By a linear transformation, or by a quadratic Tschirnhaus transformation whose coefficients involve a single square root, any equation can be reduced to a principal equation.*

119. The Bring-Jerrard normal form. We shall transform any principal equation into one in which the coefficients of y^{n-1}, y^{n-2}, y^{n-3} are all zero. By hypothesis, $s_1 = s_2 = 0$ in (1). If also $s_3 = 0$, then $c_1 = c_2 = c_3 = 0$ by Newton's identities, and no transformation is necessary. Next, let $s_3 \neq 0$. The functions

(3) $\quad g(x) = x^3 + ax^2 - s_3/n, \quad h(x) = x^4 + bx^2 - s_4/n$

evidently have the property $\sum g = 0$, $\sum h = 0$, where $\sum g$ denotes $\sum g(x_k)$. The conditions for $\sum xg = 0$, $\sum xh = 0$ are $s_4 + as_3 = 0$, $s_5 + bs_3 = 0$, and may be satisfied by choice of a and b. The function

(4) $\quad\quad\quad\quad\quad \psi(x) = zg + wh$

[1] Tschirnhaus, Acta Eruditorum, 2, 1683, 204-7.

will have the property $\sum \psi^2 = 0$ if

(5) $\qquad z^2 \sum g^2 + 2zw \sum gh + w^2 \sum h^2 = 0.$

This homogeneous quadratic equation in z and w can evidently be satisfied by values of z and w not both zero. Then x and ψ have the properties

(6) $\qquad \sum x = 0, \sum x^2 = 0, \sum \psi = 0, \sum \psi^2 = 0, \sum x\psi = 0.$

Write $y = ux + v\psi$. Then $\sum y = 0$, $\sum y^2 = 0$ for every u and v. The condition for $\sum y^3 = 0$ is a cubic equation in $u:v$ and hence can be satisfied by values of u and v not both zero. In the resulting equation in y, the coefficients of y^{n-1}, y^{n-2}, y^{n-3} are therefore all zero.

We saw that any equation $F(z) = 0$ can be reduced to a principal equation $f(x) = 0$ by a transformation $x = P(z)$. To $f(x) = 0$ we just applied a transformation $y = Q(x)$. Hence the final equation in y may be derived from $F(z) = 0$ by the single transformation $y = Q[P(z)]$. In other words, *the product of two Tschirnhaus transformations is a Tschirnhaus transformation.*

This completes the proof of

THEOREM 2. *By means of a Tschirnhaus transformation whose coefficients involve a cube root and three square roots, any equation of degree n can be transformed into an equation of degree n in y in which the coefficients of y^{n-1}, y^{n-2}, y^{n-3} are all zero.*

This theorem is usually ascribed to Jerrard.[1] But for $n = 5$ it was obtained much earlier by E. S. Bring.[2] We therefore call

---

[1] Math. Researches, Bristol and London, 1834, Pt. II. Report by W. R. Hamilton, British Assoc. Report, 5, 1837, 295–348.

[2] Meletemata . . . transformationem aequationum algebraicarum, Lund, 1786. Substance reproduced in Quar. Jour. Math., 6, 1864, 45–47; Archiv. Math. Phys., 41, 1864, 105–112; Annali di Mat., 6, 1864, 33–42.

(7) $$y^5 + dy + e = 0$$

the Bring-Jerrard normal form of quintic equations.

EXERCISES

1. Any equation of degree 3 or 4 can be solved by radicals.

2. Every quintic equation can be reduced to the form $y^5 + dy^2 + e = 0$ by a Tschirnhaus transformation whose coefficients involve a square root and a root of a quartic equation. Hint: If $c_1 = c_2 = 0$, then $s_4 + 4c_4 = 0$; hence use $\Sigma y^4 = 0$.

3. By choice of t in $y = tz$, reduce (7) and the quintic in Ex. 2 to normal forms involving a single parameter.

4. Derive (7) geometrically. Hints: Let the transformation $y = a_0 + a_1 x + a_2 x^2 + a_3 x^3 + a_4 x^4$ replace (1) by $y^5 + d_1 y^4 + d_2 y^3 + \cdots = 0$. Then d_j is a homogeneous polynomial in $a_0, \ldots, a_4$ of degree j. By means of $d_1 = 0$, express a_0 as a linear homogeneous function of $a_1, \ldots, a_4$. Interpret the latter as the four homogeneous coordinates of a point P in space (the lengths of the perpendiculars from P to the faces of a fixed tetrahedron of reference being proportional to $a_1, \ldots, a_4$). Eliminate a_0 from $d_2 = 0$ and $d_3 = 0$. The resulting quadratic equation represents a quadric surface, which contains real or imaginary straight lines; usually there are two such lines through each point of the surface and then they are found by means of a quadratic equation in one unknown. The intersections of a chosen one of these lines with the cubic surface obtained from $d_3 = 0$ are found by solving a cubic equation in one unknown.

120. The Brioschi normal form of quintic equations. In the preceding Ex. 3, we saw that any quintic equation can be reduced to a normal form involving a single parameter by means of transformations involving both square roots and a cube root. By means of a transformation involving only square roots, we shall now reduce any sufficiently general quintic equation to a remarkable form which also involves a single parameter and which plays a central rôle in the theory of quintic equations.

We start with a principal quintic $f(x) = 0$, having therefore $s_1 = 0$, $s_2 = 0$. We assume that $s_3 \neq 0$, and employ the polynomial (4) having the properties (6). We shall first prove the existence of constants $p, q, \ldots, t$ (p and q not both zero) such that

(8) \quad $p\psi^2 + 2qx\psi + rx^2 - a\psi - bx + t \equiv 0 \quad [\bmod f(x)],$

which means that the left member is the product of $f(x)$ by a polynomial in x.

First, let $z = 1, w = 0$ in (4), whence ψ is the cubic g. From ψ^2 we eliminate x^6 and x^5 by means of $f(x) = 0$, then eliminate x^4 by means of $x\psi$ and finally x^3 by means of ψ; we get (8) with $p = 1$.

Second, let $w \neq 0$. We may take $w = 1$ and write

$$\psi = x^4 + zx^3 + dx^2 + e.$$

For a certain quadratic function Q of x, we have

$$C \equiv x\psi - f(x) - z\psi = kx^3 + Q, \quad k = d - z^2.$$

If $k = 0$, this gives (8) with $p = 0, 2q = 1$. If $k \neq 0$, we eliminate x^5 and higher powers of x from ψ^2 by means of $f(x) = 0$, then x^4 by means of ψ, and finally x^3 by means of C, and obtain (8) with $p = 1$.

Inserting the five roots of $f(x) = 0$ into (8), summing, and applying (6), we see that $t = 0$.

If χ and ϕ are any linear functions of x and ψ, (6) imply

(9) \quad $\sum \chi = 0, \sum \chi^2 = 0, \sum \phi = 0, \sum \phi^2 = 0, \sum \chi\phi = 0,$

while (8) implies a similar relation

(10) \quad $p\chi^2 + 2q\phi\chi + r\phi^2 - a\chi - b\phi \equiv 0 \, [\bmod f(x)].$

We employ the identity

(11) $$mF \equiv (d\chi + e\phi)^2 - c(a\chi + b\phi)^2,$$
$$F = p\chi^2 + 2q\phi\chi + r\phi^2,$$

where

(12) $$m = pb^2 - 2qab + ra^2, \quad c = q^2 - pr, \quad d = bp - aq,$$
$$e = bq - ar.$$

We exclude the special quintics for which $c = 0$ and those for which both a and b are zero. The linear functions χ and ϕ of x and ψ may be chosen so that $a\chi + b\phi$ is identical with $ux + v\psi$, where u and v have any prescribed values not both zero. We can choose them so that $ux + v\psi$ vanishes for no root of $f(x) = 0$. Hence it is permissible to apply the transformation

(13) $$y = \frac{d\chi + e\phi}{a\chi + b\phi}$$

to $f(x) = 0$. Denote the resulting equation by

(14) $$y^5 + a_1 y^4 + a_2 y^3 + a_3 y^2 + a_4 y + a_5 = 0.$$

We determine its coefficients as follows: By (13),

$$y + c^{\frac{1}{2}} = \frac{d\chi + e\phi + c^{\frac{1}{2}}(a\chi + b\phi)}{a\chi + b\phi}.$$

If m in (12) were zero, F would be divisible by $a\chi + b\phi$ and the quotient of (10) by $a\chi + b\phi$ would be divisible by $f(x)$, whereas its degree is ≤ 4. Then if the numerator of the preceding fraction is zero for either value of $c^{\frac{1}{2}}$, (11) shows that $F = 0$, and thus (10) shows that $a\chi + b\phi$ is divisible by $f(x)$, contrary to its degree.

By (10) and (11),
$$m(a\chi + b\phi) \equiv (d\chi + e\phi)^2 - c(a\chi + b\phi)^2 \quad [\operatorname{mod} f(x)].$$

Hence, by division,
$$\frac{m}{y + c^{\frac{1}{2}}} \equiv d\chi + e\phi - c^{\frac{1}{2}}(a\chi + b\phi) \quad [\operatorname{mod} f(x)].$$

By (9), this implies
$$\sum z = 0, \quad \sum z^2 = 0 \quad \text{if} \quad z = 1/(y + c^{\frac{1}{2}}).$$

Hence if we replace y by $(1 - zc^{\frac{1}{2}})/z$ in (14), we obtain
$$(1 - zc^{\frac{1}{2}})^5 + a_1 z (1 - zc^{\frac{1}{2}})^4 + a_2 z^2 (1 - zc^{\frac{1}{2}})^3$$
$$+ a_3 z^3 (1 - zc^{\frac{1}{2}})^2 + a_4 z^4 (1 - zc^{\frac{1}{2}}) + a_5 z^5 = 0,$$

in which the coefficients of z^4 and z^3 must vanish. Thus
$$5c^2 - 4a_1 c^{\frac{3}{2}} + 3a_2 c - 2a_3 c^{\frac{1}{2}} + a_4 = 0,$$
$$-10c^{\frac{3}{2}} + 6a_1 c - 3a_2 c^{\frac{1}{2}} + a_3 = 0.$$

Since these hold for both values of $c^{\frac{1}{2}}$,
$$5c^2 + 3a_2 c + a_4 = 0, \quad 4a_1 c + 2a_3 = 0,$$
$$10c + 3a_2 = 0, \quad 6a_1 c + a_3 = 0,$$

whence $a_1 = a_3 = 0$, $a_2 = -\frac{1}{3} 10c$, $a_4 = 5c^2$. Write $c = -3C$. Then (14) becomes

(15) $\qquad y^5 + 10Cy^3 + 45C^2 y + a_5 = 0.$

This Brioschi[1] normal form is especially adapted for solution[2] in terms of elliptic functions. The effect of writing $y = C^{\frac{1}{2}} Y$ in

---

[1] Annali di Mat., 1, 1858, 256–9, 326–8.
[2] Kiepert, Jour. für Math., 87, 1879, 130.

(15) is to put $C = 1$. However, by a transformation not involving radicals, we may pass from (15) to a quintic equation having a single parameter (§131, end).

The only irrationality employed in the transformation (13) of the principal quintic $f(x) = 0$ to the form (15) was the square root of the discriminant of (5). We may express the latter in terms of the discriminant D^2 of $f(x)$. By definition, D is a determinant whose kth row is 1, x_k, x^2_k, x^3_k, x^4_k. In view of (3), D is equal to the determinant whose kth row is

$$1, \ x_k, \ x^2_k, \ g(x_k), \ h(x_k).$$

Let D' denote the determinant having these elements in the kth column. Forming the product $D'D$ by using the rows of D' and the columns of D, we see that

$$D^2 = \begin{vmatrix} 5 & 0 & 0 & 0 & 0 \\ 0 & 0 & s_3 & 0 & 0 \\ 0 & s_3 & s_4 & \sum x^2 g & \sum x^2 h \\ 0 & 0 & \sum x^2 g & \sum g^2 & \sum gh \\ 0 & 0 & \sum x^2 h & \sum gh & \sum h^2 \end{vmatrix} = -5s_3^2 \begin{vmatrix} \sum g^2 & \sum gh \\ \sum gh & \sum h^2 \end{vmatrix}.$$

Hence the discriminant of equation (5) is equal to $\frac{1}{5}D^2/s_3^2$.

THEOREM[1] 3. *If we denote the product of the squares of the differences of the roots of the principal quintic $f(x) = 0$ by $5^5\Delta$, that equation can be reduced to the form (15) by a Tschirnhaus transformation whose coefficients involve the single irrationality $\Delta^{\frac{1}{2}}$.*

Brioschi[2] proved the existence of $\frac{1}{2}(n-1)$ polynomials ϕ, χ_i $[i = 1, \ldots, \frac{1}{2}(n-3)]$ such that the sum, sum of squares, and

---

[1] The above proof is a material simplification of that by Gordan, Math. Annalen, 28, 1887, 152–166 (Jour. de Math., (4), 1, 1885, 455–8). He introduced an unnecessary irrationality, the square root of the discriminant of F, since he used instead of (13) a transformation involving the linear factors of F. The part of the proof beginning with (11) follows Weber, Algebra, ed. 2, 1898, I, 264–7.

[2] Rendiconti Ist. Lombardo, (2), 20, 1887, 364–70; Opere Mat., III, 293–9.

sum of products of any two, summed for all the roots of the equation of degree n, are all zero. This was proved in (9) above when $n = 5$.

In his Chicago thesis to be published soon, R. Garver proves that every Tschirnhaus transformation of a sufficiently general principal equation of degree n into another principal equation can be expressed in the form $y = u\chi + v\phi$, where χ and ϕ are polynomials of degrees $n - 1$ and $n - 2$ which satisfy (9).

Chapter XIII

GROUPS OF THE REGULAR SOLIDS; QUINTIC EQUATIONS

121. Introduction. The first part of this chapter gives a theory of the invariants of each finite group of linear transformations on two variables. These groups are isomorphic to groups of rotations leaving unaltered the various regular solids. The final part is an application to the resolvents of the general equation of the fifth degree and furnishes a method of solving the latter. The subject plays an important rôle in the theories of elliptic modular functions and automorphic functions.

The theory is due to Klein, whose book[1] is a classic, but presents difficulties to beginners on account of the introduction of ideas from many branches of mathematics. We shall give a simple exposition of the essentials of this interesting theory.

122. Linear fractional transformation on z corresponding to a rotation. To readers acquainted with the theory of functions of a complex variable there is available[2] a short proof of the desired formula (7). We shall give here a strictly elementary proof.

We first define the stereographic projection of a sphere upon a plane. Employ a rectangular coordinate system in space. The sphere with radius unity and center at the origin O has the equation

(1) $$\xi^2 + \eta^2 + \zeta^2 = 1.$$

From the point $N = (0, 0, 1)$ project an arbitrary point $P = (\xi, \eta, \zeta)$ of the sphere to a point $Q = (x, y)$ of the $\xi\eta$-plane.

---

[1] Vorlesungen über das Ikosaeder, 1884.

[2] Burkhardt's Funktiontheoretische Vorlesungen, I, 2, 1903, 49–52 (English transl. by Rasor, 81).

Let T be the fourth vertex of the rectangle having sides ON and OQ. Let SPF be parallel to NO. Let FG and QH be parallel to the η-axis. Then $OG = \xi$, $FG = \eta$, $PF = \zeta$, $OH = x$, $QH = y$. Since the triangles NSP and NTQ are similar,

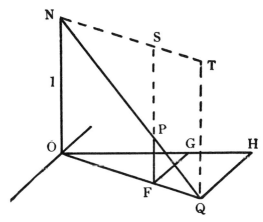

Fig. 1.

$1 - \zeta : 1 = SP : TQ = NS : NT = OF : OQ = \eta : y = \xi : x.$

Hence

(2) $$x = \frac{\xi}{1-\zeta}, \quad y = \frac{\eta}{1-\zeta}.$$

By (1),

$$1 + x^2 + y^2 = \frac{(1-\zeta)^2 + \xi^2 + \eta^2}{(1-\zeta)^2} = \frac{1 - 2\zeta + 1}{(1-\zeta)^2} = \frac{2}{1-\zeta},$$

whence

(3) $\xi = 2x/r, \quad \eta = 2y/r, \quad \zeta = 1 - 2/r, \quad r = 1 + x^2 + y^2.$

Write $z = x + iy$. Formulas (2) and (3) establish a one-to-one correspondence between the points of the unit sphere and the points of the complex z-plane, provided we agree to identify all points at infinity in the z-plane (so that $z = \infty$ alone corresponds to N).

Let $E = (\lambda, \mu, \nu)$ and $P = (\xi, \eta, \zeta)$ be any points on the sphere having radius unity and center at the origin O. Write

(4) $\qquad a = \lambda \sin \dfrac{\alpha}{2}, \quad b = \mu \sin \dfrac{\alpha}{2}, \quad c = \nu \sin \dfrac{\alpha}{2}, \quad d = \cos \dfrac{\alpha}{2},$

whence

(5) $\qquad\qquad\qquad a^2 + b^2 + c^2 + d^2 = 1.$

Consider the rotation about an axis OE through angle α counter-clockwise when viewed from E toward O. Let it replace (ξ, η, ζ) by (ξ', η', ζ'). We employ the formulas (28) of §46 which represent the rotation, with a_1, a_2, a_3 replaced by a, b, c, and w_1, w_2, w_3 replaced by ξ, η, ζ, and p_1, p_2, p_3 replaced by ξ', η', ζ'.

Write $z = x + iy$, $w = x - iy$. Then (3) become

$$\xi = \frac{z+w}{1+zw}, \quad \eta = \frac{i(w-z)}{1+zw}, \quad \zeta = \frac{zw-1}{1+zw}.$$

Hence

$$\tfrac{1}{2}(1+zw)(1-\zeta') = c^2 + d^2 + zw(1 - c^2 - d^2) + ez + \bar{e}w,$$

where $e = bd - ac + i(ad + bc) = (b + ia)(d + ic)$. Replacing $1 - c^2 - d^2$ by the equal value $a^2 + b^2$, we get

$$\tfrac{1}{2}(1+zw)(1-\zeta') = [(b+ia)z + d - ic]f,$$
$$f = (b-ia)w + d + ic.$$

Similarly,

$$\tfrac{1}{2}(1+zw)(\xi' + i\eta') = [(d+ic)z - (b-ia)]f.$$

Writing $z' = x' + iy'$, and employing (2) in accents, we get

(6) $$z = \frac{\xi + i\eta}{1 - \zeta}, \quad z' = \frac{\xi' + i\eta'}{1 - \zeta'},$$

(7) $$z' = \frac{(d + ic)z - (b - ia)}{(b + ia)z + d - ic}.$$

THEOREM 1. *Every rotation corresponds to a transformation* (7).

EXERCISES

1. By (4) and (7), with $\lambda = \mu = 0$, $\nu = 1$, prove that the rotation about the ζ-axis ON through angle α counterclockwise viewed from N is represented by

(8) $$z' = e^{i\alpha} z.$$

2. Prove (8) by use of the formulas from analytic geometry,

$$\xi' = \xi \cos \alpha - \eta \sin \alpha, \quad \eta' = \xi \sin \alpha + \eta \cos \alpha.$$

123. The tetrahedral group. Consider a cube inscribed in a sphere with the center O and radius unity. Four of its vertices 1, 2, 3, 4, shown in Fig. 2, are the vertices of a regular tetrahedron. Their diametral points $1'$, $2'$, $3'$, $4'$ are the vertices of a second regular tetrahedron called the *diametral* tetrahedron to 1234.

The mutually perpendicular lines AA', BB', CC', each of which joins the mid points of a pair of opposite edges of 1234, are taken as the ξ-axis, η-axis, ζ-axis, respectively, their positive directions being from O to A, B, C, respectively.

The tetrahedron 1234 is unaltered by a rotation through angle $0°$, $120°$, or $240°$ about any of the lines $O1$, $O2$, $O3$, $O4$ as an axis, and also by a rotation through $180°$ about one of the axes AA', BB', CC'. These include the identity rotation (through angle $0°$), eight rotations of order 3, and three of order 2. They give all

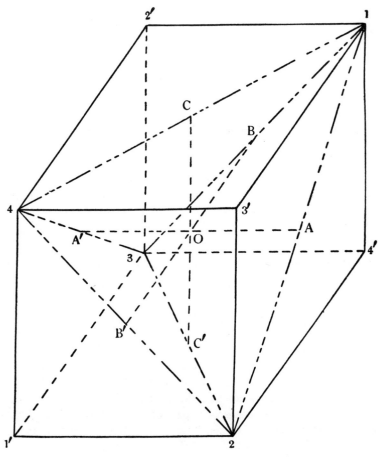

Fig. 2.

the rotations leaving the tetrahedron 1234 unaltered and hence form a group called the *tetrahedral group* R_{12}. For, exactly three such rotations leave the vertex 1 unaltered, while their products by any one rotation (about axis $O4$) which carries 1 to 2 give all the rotations which carry 1 to 2, and similarly when we use 3 or 4 instead of 2, whence the total number is 3×4.

Consider in particular the rotation S about the axis $O1$ through angle 120° counterclockwise when viewed from 1, and the rotation R about the axis CC' through angle 180°. Since R interchanges 1 and 4, RSR and RS^2R leave 4 unaltered and hence give rotations about the axis $O4$ which carry 1 to 3 and 2 respectively. By the final remark in the preceding paragraph, the identity, S, and S^2, together with their products by RS^2R, RSR, and R (which carry 1 to 2, 3, and 4 respectively) give all the rotations of R_{12}.

THEOREM 2. *The tetrahedral group R_{12} of all rotations which leave unaltered the regular tetrahedron 1234 is generated by the two rotations S and R.*

We shall next obtain by §122 the linear fractional transformations on z which correspond to the rotations S and R. In the case of R, we have merely to employ (8) for $\alpha = \pi$ to obtain $z' = -z$. Next, the axis $O1$ of rotation S makes the same acute angle with the positive directions of the coordinate axes and hence the coordinates of point 1 are $\lambda = \mu = \nu = 3^{-\frac{1}{2}}$. Since $\alpha = 120°$ and $\cos 60° = \frac{1}{2}$, $\sin 60° = \frac{1}{2} 3^{\frac{1}{2}}$, (4) gives $a = b = c = d = \frac{1}{2}$. Then (7) becomes

(9) $$z' = \frac{z+i}{z-i}.$$

THEOREM 3. *The group F_{12} of all tetrahedral linear fractional transformations is generated by (9) and $z' = -z$.*

To find the product AB of a transformation A defined by (7) by a second transformation $B : z' = f(z)$, we write the latter in the form $z'' = f(z')$ and eliminate z' between the final equation and (7). The resulting transformation $z'' = \phi(z)$ is defined to be AB.

EXERCISES

1. The group R_{12} is simply isomorphic to the alternating group on the letters 1, 2, 3, 4 denoting the vertices.

2. The 12 transformations of F_{12} are

(10) $\qquad z' = \pm z, \pm \dfrac{1}{z}, \pm i\dfrac{z+1}{z-1}, \pm i\dfrac{z-1}{z+1}, \pm \dfrac{z+i}{z-i}, \pm \dfrac{z-i}{z+i}.$

Hints: $S^2 = S^{-1}$ is given by the third function (10) with the upper sign, while $L \equiv S^{-1}RS$ is $z' = 1/z$. The last may be computed as a product or proved by noting that L leaves A fixed and hence is the rotation through $180°$ about axis AA'. For $\lambda = 1$, $\mu = \nu = 0$, $\alpha = 180°$, (4) gives $a = 1$, $b = c = d = 0$, whence (7) becomes $z' = i/(iz)$.

3. If R (or $z' = -z$) is the rotation through $180°$ about axis CC' and if L (or $z' = 1/z$) is that about AA', their product $M = RL$ (or $z' = -1/z$) is the rotation through $180°$ about the common perpendicular BB' to CC' and AA'. Hence show that R, L, M, and the identity rotation form a commutative group of order 4.

To each transformation (7), of determinant unity, corresponds the linear homogeneous transformation

(11) $\qquad \begin{aligned} U &= (d+ic)u - (b-ia)v, \\ V &= (b+ia)u + (d-ic)v, \end{aligned}$

of determinant unity, as well as that obtained by changing the signs of a, b, c, d, which is therefore the product of (11) and $U = -u$, $V = -v$, in either order.

In particular, to $z' = iz/(-i)$ and (9) correspond

(12) $\qquad\qquad\qquad U = iu, V = -iv,$

(13) $\qquad \begin{aligned} U &= \tfrac{1}{2}(1+i)u - \tfrac{1}{2}(1-i)v, \\ V &= \tfrac{1}{2}(1+i)u + \tfrac{1}{2}(1-i)v. \end{aligned}$

§ 124] INVARIANTS OF TETRAHEDRAL GROUP

THEOREM 4. *The group H_{24} of all tetrahedral homogeneous linear transformations is generated by* (12) *and* (13), *each of determinant unity.*

124. Invariants of the tetrahedral group H_{24}.

The vertices of the tetrahedron 1234 have the coordinates

(14) $\quad\quad\quad \xi = \pm 3^{-\frac{1}{2}}, \ \eta = \pm 3^{-\frac{1}{2}}, \ \zeta = \pm 3^{-\frac{1}{2}},$

where an odd number of the signs are plus. By (6), the corresponding values of z are $\pm a$ and $\pm b$, where

$$a = \frac{1+i}{3^{\frac{1}{2}} - 1}, \quad b = \frac{1-i}{3^{\frac{1}{2}} + 1}.$$

We seek the product Φ of the factors $u \mp av$, $u \mp bv$. Since $a^2 = i(2 + 3^{\frac{1}{2}})$ and $b^2 = -i(2 - 3^{\frac{1}{2}})$ have the product 1 and sum $2 \cdot 3^{\frac{1}{2}} i$, we get

(15) $\quad\quad\quad \Phi = u^4 - 2 \cdot 3^{\frac{1}{2}} i u^2 v^2 + v^4.$

Since the 12 rotations which leave the tetrahedron unaltered merely permute the four vertices, each of the 24 tetrahedral homogeneous transformations merely permutes the above four linear functions, apart from constant factors, and hence leaves Φ unaltered up to a constant factor. That factor is evidently 1 for (12), while for (13) it is

$$[\tfrac{1}{2}(1+i)]^4 (1 - 2 \cdot 3^{\frac{1}{2}} i + 1) = \tfrac{1}{2}(-1 + 3^{\frac{1}{2}} i) = e^{2\pi i/3}.$$

Hence Φ^3 is an absolute invariant of the group H_{24}.

The vertices of the diametral tetrahedron $1'2'3'4'$ have the coordinates (14), where now an odd number of the signs are minus. The corresponding values (6) of z are $\pm \bar{a}$ and $\pm \bar{b}$. Hence the group has the absolute invariants $\Phi\Psi$ and Ψ^3, where

(16) $\quad\quad\quad \Psi = u^4 + 2 \cdot 3^{\frac{1}{2}} i u^2 v^2 + v^4.$

Any rotation which leaves the tetrahedron 1234 unaltered merely permutes its six edges and hence their middle points. It therefore permutes the points in which the axes of coordinates intersect the unit sphere; the corresponding values of z are evidently 0, ∞, ± 1, $\pm i$. The product of the linear functions which vanish for these values of $z = u/v$ is

$$(17) \qquad t = uv(u^4 - v^4).$$

It is an absolute invariant of the group, being unaltered by (12) and (13).

A point on the sphere is called a *special* point with respect to the group R_{12} of rotations leaving the tetrahedron 1234 unaltered if the point takes fewer than 12 distinct positions under these 12 rotations. In other words, a point is a special point if and only if it is unaltered by at least one of those rotations other than the identity, and then the point lies on the axis of the rotation.

The axes AA', BB', CC' of the rotations of order 2 are the coordinate axes. They intersect the sphere in the six special points for which the invariant t vanishes.

The eight rotations of order 3 of R_{12} have as axes the diameters through the vertices. The latter are the four special points for which the invariant Φ vanishes. The remaining intersections of those diameters with the sphere are the vertices of the diametral tetrahedron and hence are the four special points for which the invariant Ψ vanishes.

Consider any homogeneous polynomial in u and v which is an invariant of H_{24}. If it vanishes for a special point, it has one of the factors t, Φ, Ψ. After removing all such factors, we obtain an invariant quotient Q which vanishes for no special point. Since $Q(u, v)$ is homogeneous in u and v, the equation $Q(z, 1) = 0$ has a root z', so that $Q(u, v)$ vanishes for $u = U$, $v = V$, where $U/V = z'$. Let f and g denote any two of the absolute invariants t^2, Φ^3, Ψ^3, each of degree 12. Then also $f - cg$ is an absolute invariant, if c is any constant. The condition that it shall vanish at

U, V uniquely determines c, since z' is not a special point and therefore $g(U, V) \neq 0$. Hence Q has the factor $f - cg$. The quotient is either a constant or has another such factor. Proceeding similarly, we have

THEOREM 5. *Every homogeneous polynomial in u and v which is invariant under the 24 tetrahedral homogeneous linear transformations is a product of factors t, Φ, Ψ, $f - cg$, where the c's are constants $\neq 0$, while f and g denote any two of the absolute invariants t^2, Φ^3, Ψ^3 of degree 12.*

Hence an invariant of degree 12 which vanishes at no one of the $6 + 4 + 4$ special points can be expressed in each of the forms $a(t^2 - c\Phi^3)$, $b(t^2 - d\Psi^3)$, where a, b, c, d are constants $\neq 0$. Thus t^2, Φ^3, and Ψ^3 satisfy a linear identity. By considering the terms u^{12} and $u^{11}v$, we find

(18) $$12(-3)^{\frac{1}{2}} t^2 + \Phi^3 - \Psi^3 \equiv 0.$$

EXERCISES

1. If z and z' are the stereographic projections of two diametral points on the unit sphere, $\bar{z}z' = -1$. Hence derive (16) from (15) by replacing z by $-1/z$ in $\bar{\Phi}(u/v) = 0$.

2. The Hessian of Φ is $-48 \cdot 3^{\frac{1}{2}} i \Psi$. The Jacobian of Ψ and Φ is $32 \cdot 3^{\frac{1}{2}} it$. When Φ is taken as the fundamental quartic form, the invariant I is zero, while $J = -4/(3 \cdot 3^{\frac{1}{2}} i)$. The syzygy (22) of §12 reduces to (18).

125. The octahedral group. In Fig. 2, the solid composed of the two pyramids with the vertices C and C' and common square base $ABA'B'$ is a regular octahedron. It is unaltered by the rotations of the group R_{12} which leaves the tetrahedron 1234 unchanged, together with 12 rotations which interchange the latter and its diametral tetrahedron $1'2'3'4'$, viz., the 6 rotations through $\pm 90°$ about the axes AA', BB', and CC', and the 6 rotations through $180°$ about the 6 diameters each bisecting two edges of the octahedron. Products of these rotations replace A by A', B, B', C, C'; since only 4 rotations about OA leave the octahedron unaltered, there are only 6×4 rotations leaving it unaltered. Hence all rotations leaving the octahedron unaltered form a group R_{24}. It is generated by R and S of §123 and the rotation E about the ζ-axis through $90°$ counterclockwise when viewed from C. By (8), E is represented by $z' = iz$, one of whose two homogeneous forms is

(19) $\qquad U = 2^{-\frac{1}{2}}(1 + i)u, \qquad V = 2^{-\frac{1}{2}}(1 - i)v.$

THEOREM 6. *All rotations leaving unaltered a regular octahedron form a group R_{24} of order 24. The octahedron group H_{48} of linear homogeneous transformations is generated by* (19) *and the two generators* (12) *and* (13) *of the tetrahedral subgroup H_{24}.*

126. The invariants of the octahedral group H_{48}. Consider the octahedron whose vertices are the intersections of the three coordinate axes with the unit sphere.

Since the function t given by (17) vanishes only for the six values of z which correspond to the vertices, t is an invariant of H_{48}. We saw that t is unaltered by all the transformations of H_{24}. Under transformation (19), t becomes $-t$. Hence t^2 is an absolute invariant of H_{48}.

The Hessian of t is $-25W$, where

(20) $\qquad W = u^8 + 14u^4 v^4 + v^8 = \Phi \Psi.$

The Jacobian of t and W is the product of

(21) $\qquad \chi = u^{12} - 33u^8 v^4 - 33u^4 v^8 + v^{12}$

by a constant. Since each transformation of H_{48} is of determinant unity and replaces t by $\pm t$, the properties of the Hessian and the Jacobian (§§3, 5) show that the transformation leaves W unaltered and replaces χ by $\pm \chi$.

A point on the unit sphere is called a *special* point if it is unaltered by at least one rotation, other than the identity, of R_{24} and hence lies on the axis of that rotation. We saw that the special points on axes of rotations of the tetrahedral group R_{12} are the points for which t or $W = \Phi \Psi$ vanishes. By §125, the only further special points are the central projections on the sphere of the mid points of the 12 edges of the octahedron. For these points χ must vanish. In fact, since χ is an invariant it vanishes for some set of 12 special points which is distinct from the set of the six vertices each counted twice, since $\chi \neq t^2$ (cf. Ex. 3 below). The discussion at the end of §124 therefore leads to the

THEOREM 7. *Every homogeneous polynomial in u and v which is invariant under the group of 48 octahedral linear homogeneous transformations is a product of factors t, W, χ, $f - cg$, where the c's are constants, while f and g denote any two of the absolute invariants t^4, W^3, χ^2 of degree 24. The latter satisfy the linear identity*

(22) $$108 t^4 - W^3 + \chi^2 \equiv 0.$$

EXERCISES

1. Why is $\Phi \Psi$ invariant under H_{48}?

2. R_{24} is simply isomorphic to the symmetric group on 4 letters representing the four diagonals $11', \ldots, 44'$ of the cube.

3. The diameter which bisects the edge joining $(0, 0, 1)$ and $(1, 0, 0)$ meets the unit sphere at $(\frac{1}{2}2^{\frac{1}{2}}, 0, \frac{1}{2}2^{\frac{1}{2}})$, whose z is $1 + 2^{\frac{1}{2}}$. For this value of u and for $v = 1$, $\chi = 0$. The invariant χ therefore vanishes at the central projections on the sphere of the mid points of all 12 edges.

4. There exists a fundamental system of invariants of any finite group G of linear transformations. By Hilbert's theorem (§17) every invariant I of G is expressible in the form $I \equiv E_1 I_1 + \cdots + E_m I_m$, where the I_j are invariants. Apply the transformations of G and add. Why does this proof hold also for relative invariants?

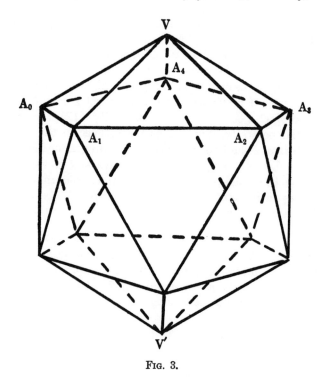

Fig. 3.

127. The icosahedral group. Consider a regular icosahedron I inscribed in a sphere having the center O and radius unity. It has 20 triangular faces, 12 vertices, and 30 edges. If V is any vertex, I is unaltered by the rotation S about the axis OV through angle $2\pi/5$ counterclockwise viewed from V. Letter the triangular faces $A_0 V A_1, A_1 V A_2, \ldots, A_4 V A_0$ having the common vertex V so that S carries A_0 to A_1, A_1 to A_2, ..., A_4 to A_0. The rotation through angle π about the axis joining O to the mid point of an edge AB will be denoted temporarily by $[AB]$; it leaves I unaltered. Write T for $[A_0 V]$. Since S carries the mid point of $A_0 V$ to that of $A_1 V$, $S^{-1} T S$ leaves the latter point unaltered and hence is $[A_1 V]$. Similarly, $S^{-j} TS^j = [A_j V]$. We shall speak of these five rotations $[A_j V]$ of order 2 and the five powers of S as the ten rotations connected with vertex V. They are transformed into the ten rotations connected with vertex A_k by $[A_k V]$, which replaces V by A_k. A glance at Fig. 3 shows that the vertex V', diametrically opposite to V, is the only vertex which cannot be reached by an edge starting from one of the vertices $A_0, \ldots, A_4$. If $A_0 B$ is either of the two edges having A_0 as one end point and distinct from $A_0 A_4$, $A_0 V$, $A_0 A_1$, the rotation $[A_0 B]$ transforms the five rotations about the axis OA_0 into the five rotations about the axis OB, and one of the latter carries A_0 to V'. We have now proved that suitably chosen products of S and T carry V to all 12 vertices. The five powers of S are the only rotations of I into itself which leave V unaltered. Hence there exist exactly 5×12 rotations of I into itself.

THEOREM 8. *S and T generate the icosahedral group R_{60} of all rotations of I into itself.*

EXERCISES

1. Since T interchanges V and A_0, as well as A_1 and A_4, the product $ST = (VA_0 A_4) \cdots$ is a rotation of order 3 whose axis joins O to the mid point of face $VA_0 A_4$.

2. The proof that S and T generate the ten rotations connected with any of the five vertices A_k shows similarly that they generate the ten rotations connected with any of the five vertices of type B and the ten connected with V'.

3. Hence S and T generate the 6×4 rotations of order 5 about the 6 diameters joining pairs of opposite vertices, the 15 rotations of order 2 about diameters each of which bisects two edges, and (Ex. 1) the 2×10 rotations of order 3 about diameters each of which passes through the mid points of two faces. These with the identity rotation give the 60 rotations of R_{60}.

128. The linear homogeneous icosahedral group H_{120}. Choose rectangular coordinate axes so that the positive ζ-axis extends from O to V and the axis of the rotation T lies in the second and fourth quadrants of the $\xi\zeta$-plane. By (8), rotation S is $z' = \epsilon z$, one of whose two homogeneous forms is

(23) $$U = \epsilon^3 u, \quad V = \epsilon^2 v, \quad \epsilon = e^{2\pi i/5}.$$

The vertex A_0 is represented by a real negative number n. Since S carries A_i to A_{i+1}, the vertices A_1, A_2, A_3, A_4 are represented by $\epsilon n, \epsilon^2 n, \epsilon^3 n, \epsilon^4 n$. Since T interchanges A_0 and V, which are represented by n and ∞, and interchanges the diametrically opposite points, represented by $m = -n^{-1}$ and 0, we see that

$$T: \quad z' = \frac{nz + 1}{z - n}.$$

Since T interchanges A_1 and A_4, $z = \epsilon n$ implies $z' = \epsilon^4 n$. Hence $n^2(1 - \epsilon - \epsilon^4) = 1$, or $n^2(\epsilon + \epsilon^4)^2 = 1$. But $\epsilon + \epsilon^4$ is positive and n is negative. Hence $n(\epsilon + \epsilon^4) = -1$. Applying the powers of S ($z' = \epsilon z$) to m and n, we conclude that the 12 vertices of the icosahedron are represented by

(24) $$z = 0, \infty, \epsilon^k m, \epsilon^k n$$
$$(k = 0, 1, 2, 3, 4; \quad m = \epsilon + \epsilon^4, \quad n = \epsilon^2 + \epsilon^3).$$

Multiply the numerator and denominator of T by $\epsilon^3 - \epsilon^2$; we get

$$T: \quad z' = \frac{(\epsilon - \epsilon^4)z + \epsilon^3 - \epsilon^2}{(\epsilon^3 - \epsilon^2)z - (\epsilon - \epsilon^4)},$$

whose determinant is 5. Hence a homogeneous form of T is

(25) $$5^{\frac{1}{2}} U = (\epsilon - \epsilon^4)u + (\epsilon^3 - \epsilon^2)v,$$
$$5^{\frac{1}{2}} V = (\epsilon^3 - \epsilon^2)u - (\epsilon - \epsilon^4)v.$$

§ 129]   INVARIANTS OF ICOSAHEDRAL GROUP   235

As a check, the square of the latter is $U = -u$, $V = -v$.

THEOREM 9. *The icosahedral group H_{120} is generated by* (23) *and* (25).

129. The invariants of the icosahedral group H_{120}. The form which vanishes at the 12 values (24) of $z = u/v$ is

$$f = uv(u^5 - m^5 v^5)(u^5 - n^5 v^5).$$

We desire for $r = 5$ the sum s_r of the rth powers of the roots m and n of $x^2 + x - 1 = 0$. Multiply by x^{r-2}, insert the roots, and sum; we get

$$s_r + s_{r-1} - s_{r-2} = 0.$$

Using this recursion formula, we get

$$s_5 = -s_4 + s_3 = 2s_3 - s_2 = -3s_2 + 2s_1 = 5s_1 - 6 = -11.$$

Since $mn = -1$, we have

(26) $$f = uv(u^{10} + 11u^5 v^5 - v^{10}).$$

The Hessian of f is $121H$, and the Jacobian of f and H is $20T$, where

(27) $$H = -u^{20} - v^{20} + 228(u^{15} v^5 - u^5 v^{15}) - 494 u^{10} v^{10},$$

(28) $$T = u^{30} + v^{30} + 522(u^{25} v^5 - u^5 v^{25}) - 10005(u^{20} v^{10} + u^{10} v^{20}).$$

A special point of the sphere is defined to be a point which takes fewer than 60 positions under the group R_{60}. Hence it is a point on the axis of one of the rotations other than the identity. The enumeration of these rotations in the preceding Ex. 3 shows that every special point belongs to one of the following sets: the 12 vertices (at which f vanishes), the central projections on the sphere of the mid points of the 20 faces, and those of the mid points of the 30 edges. An aggregate of two or more sets, not necessarily distinct, never gives a set of 20 or 30 points. Hence the invariant H vanishes at the central projections on the sphere of the mid points of all 20 faces, while the invariant T vanishes at those of the mid points of all 30 edges.

Evidently f is unaltered by transformation (23). Computation shows that f is unaltered by (25), and hence is an absolute invariant of H_{120}. The same is true of H and T since the transformations have determinant unity. Proceeding as at the end of §124, we obtain

Theorem 10. *Every homogeneous polynomial in u and v which is invariant under the icosahedral group H_{120} is an absolute invariant which is a polynomial in the invariants f, H, and T. The latter satisfy the identity*

(29) $$T^2 \equiv 1728f^5 - H^3.$$

130. The form problems for the tetrahedral and octahedral groups. By §124, every polynomial absolute invariant of the homogeneous tetrahedral group H_{24} is a polynomial in the absolute invariants t, $W = \Phi\Psi$, and Φ^3, which satisfy the identity

(30) $$\Phi^3[12(-3)^{\frac{1}{2}} t^2 + \Phi^3] - W^3 \equiv 0.$$

Given arbitrary values for t, W, Φ^3, compatible with relation (30), we seek the sets of values of u, v. This is the *form problem* for H_{24}. From one set of solutions u, v, we evidently obtain 24 sets by applying to it the 24 substitutions of H_{24}. That there are exactly 24 sets of solutions is seen as follows: Write

(31) $$Z_1 = \Phi^6/W^3 \equiv \Phi^3/\Psi^3, \qquad X(u, v) = W/t,$$

§ 130] TETRAHEDRAL, OCTAHEDRAL FORM PROBLEMS 237

so that Z_1 and X are the given values of these absolute invariants. The first quotient is a function of $z = u/v$. The corresponding equation

$$(32) \quad (z^4 - 2\cdot 3^{\frac{1}{2}} iz^2 + 1)^3 - Z_1(z^4 + 2\cdot 3^{\frac{1}{2}} iz^2 + 1)^3 = 0$$

has 12 solutions z. For each such z, we may compute $X(z, 1)$ and then find two sets of values of u, v from

$$(33) \quad v^2 = X(u, v)/X(z, 1), \quad u = zv.$$

The first relation (33) follows from the fact that the degree 8 of W exceeds the degree 6 of t by 2. The form problem for H_{24} therefore reduces to the solution of equation (32) and extraction of square roots.

Equation (32) is invariant under the group F_{12} of all tetrahedral linear fractional transformations on z. Four of them (cf. Ex. 3, §123) are $z' = \pm z$, $z' = \pm 1/z$, and these leave invariant the function

$$(34) \quad Z_2 = (z^2 - 1)^2/(4z^2).$$

Employ the cube root $\omega = \frac{1}{2}(-1 + 3^{\frac{1}{2}} i)$ of unity. Then

$$Z_2 - \omega = (z^4 - 2\cdot 3^{\frac{1}{2}} iz^2 + 1)/(4z^2),$$

while $Z_2 - \omega^2$ is obtained by changing i to $-i$. Hence

$$(35) \quad \left(\frac{Z_2 - \omega}{Z_2 - \omega^2}\right)^3 = Z_1.$$

Hence (32) may be solved by extracting a cube root (35) to get Z_2, solving the quadratic equation (34) in z^2, and extracting the square root of the resulting value of z^2.

For the form problem of the octahedral group H_{48}, we are given arbitrary values of the absolute invariants W, t^2, $t\chi$, compatible with the relation (22) or

$$t^2(108t^4 - W^3) + (t\chi)^2 \equiv 0,$$

and seek the 48 sets of values of u, v. Write

(36) $\qquad Z = W^3/t^4, \qquad X(u, v) = Wt/\chi.$

Since X is of degree 2 in u, v, we again have (33). For $Z_1 = \Phi^3/\Psi^3$, we see from (30) and $W = \Phi\Psi$ that

(37) $\qquad -3 \cdot 144\, Z_1/(Z_1 - 1)^2 = Z.$

The first equation (36) is of degree 24 in the unknown z. Its solution is equivalent to the solution of the quadratic equation (37) for Z_1 and the earlier equations for finding z from Z_1.

THEOREM 11. *The form problems for the tetrahedral and octahedral groups are solvable by radicals.*

131. The form problem for the icosahedral group H_{120}. Given arbitrary values of the absolute invariants f, H, and T, compatible with relation (29), we seek the 120 sets of values of u, v. Write

(38) $\qquad Z = f^5/T^2, \qquad X(u, v) = fH/T.$

Since X is of degree 2 in u, v, we again have (33). Hence the form problem for H_{120} reduces essentially to the solution of $f^5 - ZT^2 = 0$, which after division by v^{60} becomes

(39) $\quad z^5(z^{10} + 11z^5 - 1)^5 - Z[z^{30} + 1 + 522(z^{25} - z^5)$
$\qquad\qquad\qquad\qquad - 10005(z^{20} + z^{10})]^2 = 0.$

§ 131] ICOSAHEDRAL FORM PROBLEM 239

This equation of degree 60, having the single parameter Z, is called the *icosahedral equation*. When Z is arbitrary this equation will be shown to be not solvable by radicals, in contrast with the form problems for the tetrahedral and octahedral groups. Our first step will be to reduce the solution of (39) to that of a remarkable quintic equation.

In (26) we gave the function f which vanishes for the twelve values of $z = u/v$ which represent the vertices of the icosahedron I. By inspection, f is unaltered by the transformation

(40) $\qquad U = -v, \qquad V = u; \qquad z' = -1/z.$

Hence the corresponding rotation F leaves I unaltered, so that (40) belongs to the group H_{120}. In §§127–8, we defined the rotation T through 180° about the axis OP in the $\xi\zeta$-plane, where P is the mid point of the edge $A_0 V$. By Ex. 3, §123, F is the rotation through 180° about the η-axis, and the product TF is the rotation through 180° about the axis which is a common perpendicular to $O\eta$ and OP.

To find the points unaltered by (40), take $z' = z$, whence $z^2 + 1 = 0$. To find the points unaltered by T, write (25) in the non-homogeneous form and take $z' = z$. Similarly for TF. In the resulting three equations, we replace z by u/v. Hence the pairs of points unaltered by F, T, TF are respectively those for which

$$A_0 = u^2 + v^2, \qquad B_0 = u^2 - 2(\epsilon^2 + \epsilon^3)uv - v^2,$$
$$C_0 = u^2 - 2(\epsilon + \epsilon^4)uv - v^2$$

vanish. Hence the points for which

(41) $\quad t_0 = A_0 B_0 C_0 = u^6 + 2u^5 v - 5u^4 v^2 - 5u^2 v^4 - 2uv^5 + v^6$

vanishes correspond to the vertices of a regular octahedron. Thus t_0 is the fundamental invariant of the octahedral group referred to axes different from those used in §126. By (23), we have

(42) $\quad S^k:\qquad U=\epsilon^{3k}u,\qquad V=\epsilon^{2k}v.$

This replaces A_0, B_0, C_0 by respectively

$$A_k=\epsilon^k u^2+\epsilon^{4k}v^2,\quad B_k=\epsilon^k u^2-2(\epsilon^2+\epsilon^3)uv-\epsilon^{4k}v^2,$$
$$C_k=\epsilon^k u^2-2(\epsilon+\epsilon^4)uv-\epsilon^{4k}v^2.$$

Write $t_k=A_k B_k C_k$. For T defined by (25),

$$UV=\tfrac{1}{5}l(u^2+uv-v^2),\qquad l=(\epsilon^3-\epsilon^2)(\epsilon-\epsilon^4)=5^{\frac{1}{2}}.$$

We readily verify that T replaces A_1 by mB_2, where

$$m=\tfrac{1}{5}(2\epsilon-\epsilon^2+\epsilon^3-2\epsilon^4),$$

C_1 by $(\epsilon^2-\epsilon^3)A_2$, and $C_1-B_1=-2luv$ by

$$-2(u^2+uv-v^2)=(\epsilon^2-\epsilon^3)A_2-(\epsilon^2+\epsilon^3)C_2.$$

Hence T replaces B_1 by $(\epsilon^2+\epsilon^3)C_2$. Since $m(\epsilon^2-\epsilon^3)(\epsilon^2+\epsilon^3)=1$, T replaces $t_1=A_1 B_1 C_1$ by t_2. This implies that T replaces t_4 by t_3 since A_{-k}, B_{-k}, C_{-k} are derived from A_k, B_k, C_k by replacing ϵ by ϵ^{-1}, which merely changes the signs of all coefficients of T, a change not affecting quadratic functions of u, v.

Since T permutes the end points of the axes of F and TF and leaves those of T unaltered, T changes A_0 and C_0 into their negatives and leaves B_0 unaltered. Hence, since T is of order 2,

(43) $\qquad T=(t_0)(t_1\,t_2)(t_3\,t_4),\qquad S=(t_0\,t_1\,t_2\,t_3\,t_4).$

Theorem 12. *The icosahedral group of 60 rotations or of 60 linear fractional transformations is simply isomorphic to the alternating group on five letters.*

§ 131]   ICOSAHEDRAL FORM PROBLEM   241

This may be seen geometrically. The 15 diameters each of which bisects two edges of the icosahedron fall into 5 triples each of three mutually perpendicular diameters (like the axes of F, T, TF). Each triple can be converted into all the remaining triples by rotations of R_{60}. Hence exactly 60/5 or 12 rotations of R_{60} leave a given triple unaltered. The latter form a subgroup of the octahedral group R_{24}, whose rotations of order 4 are not in R_{60}. Hence this subgroup is a tetrahedral group R_{12}. From (43) it is easy to pick out the 12 substitutions on $t_0, \ldots, t_4$ which leave t_0 unaltered and hence correspond to the subgroup R_{12} of R_{60}. For example, $STS^3 = (0)(2)(143)$, T, and $F = S^2 TS^3 TS^2 T = (0)(14)(23)$ generate the homogeneous form H_{24} of R_{12}.

By (43) every transformation of the icosahedral group H_{120} permutes $t_0, \ldots, t_4$ amongst themselves and hence leaves unaltered their elementary symmetric functions. Thus $t_0, \ldots, t_4$ are the roots of a quintic equation

$$t^5 + c_1 t^4 + c_2 t^3 + c_3 t^2 + c_4 t + c_5 = 0,$$

where c_k is a polynomial of degree $6k$ in u, v which is an invariant of H_{120}. The largest $6k$ is 30 and hence is < 60. Thus by §129, c_k is a product of factors f, H, T, whose degrees are 12, 20, 30. Since no such product is of degree 6 or 18, we have $c_1 = c_3 = 0$. Also, $c_2 = af$, $c_4 = bf^2$, $c_5 = cT$, where a, b, c are constants. Let s_k be the sum of the kth powers of the roots. Since $c_1 = c_3 = 0$, two of Newton's identities reduce to $s_2 + 2c_2 = 0$, $s_4 + c_2 s_2 + 4c_4 = 0$. Thus

(44) $$s_2 + 2af = 0, \qquad s_4 + (4b - 2a^2)f^2 = 0,$$
$$cT + t_0 t_1 t_2 t_3 t_4 = 0.$$

Since S^k in (42) replaces t_0 by t_k, we have

(45) $$t_k = \epsilon^{3k} u^6 + 2\epsilon^{2k} u^5 v - 5\epsilon^k u^4 v^2 + \ldots,$$

whence

$$t^2_k = \epsilon^k u^{12} + 4u^{11} v - 6\epsilon^{4k} u^{10} v^2 + \ldots,$$
$$t^4_k = \epsilon^{2k} u^{24} + 8\epsilon^k u^{23} v + 4u^{22} v^2 + \ldots.$$

The coefficients of $u^{11} v$, $u^{22} v^2$, u^{30} in the respective functions (44) are

$$20 + 2a = 0, \quad 20 + 4b - 2a^2 = 0, \quad c + 1 = 0,$$

whence $a = -10$, $b = 45$, $c = -1$. This proves

Theorem 13. *The octahedral forms $t_0, \ldots, t_4$ are the roots of*

(46) $$t^5 - 10ft^3 + 45f^2 t - T = 0.$$

This equation was first obtained by Brioschi (§120).

Introduce in place of the root t of (46) the new unknown $w = tf^2/T$, which is of degree zero in u and v and hence is a function of z. In (46), replace t by wT/f^2, multiply all terms by f^{10}/T^5, and write Z for f^5/T^2, in accord with (38). We get

(47) $$w^5 - 10Zw^3 + 45Z^2 w - Z^2 = 0.$$

132. The principal quintic resolvent of the icosahedral equation. We employ the functions t_k defined by (41) and (45). The Hessian of t_k is $400\, W_k$, where

(48) $$\begin{aligned}W_k = &- \epsilon^{4k} u^8 + \epsilon^{3k} u^7 v - 7\epsilon^{2k} u^6 v^2 - 7\epsilon^k u^5 v^3 \\ &+ 7\epsilon^{4k} u^3 v^5 - 7\epsilon^{3k} u^2 v^6 - \epsilon^{2k} uv^7 - \epsilon^k v^8.\end{aligned}$$

Since the generators T and S of the icosahedral group H_{120} are of determinant unity and permute $t_0, \ldots, t_4$ according to the substitutions (43), they permute their Hessians and hence $W_0, \ldots, W_4$ according to the same substitutions. Thus $\sum t_k^r W_k^s$ is unaltered by T and S and is therefore an invariant of H_{120}, and hence is a polynomial in f, H, T, whose degrees are 12, 20, 30, respectively. No linear combination of the latter with integral coefficients ≥ 0 reduces to 8, 14, 16, 22, or 28. Hence

§ 132]   PRINCIPAL ICOSAHEDRAL RESOLVENT   243

$$\sum W = 0, \ \sum tW = 0, \ \sum W^2 = 0, \ \sum tW^2 = 0, \ \sum t^2 W^2 = 0.$$

Thus $\sum Y = 0, \ \sum Y^2 = 0$, if

(49) $Y_k = \sigma W_k + \tau t_k W_k$ ($k = 0, 1, 2, 3, 4$).

Hence the latter are the roots of a principal quintic

(50) $Y^5 + 5aY^2 + 5bY + c = 0.$

By Newton's identities, we have $\sum Y^3 = -15a$, $\sum Y^4 = -20b$. An inspection of the degrees of the invariants yields the following identities except as to the prefixed numerical factors which were found by a comparison of single terms of each member:

$$\sum W^3 = -5 \cdot 24f^2, \quad \sum tW^3 = -5T,$$
$$\sum t^2 W^3 = -5 \cdot 72f^3, \quad \sum t^3 W^3 = -15fT,$$
$$\sum W^4 = 20fH, \quad \sum t^2 W^4 = -60f^2 H,$$
$$\sum t^3 W^4 = -5HT, \quad \sum t^4 W^4 = -540f^3 H.$$

Also $\sum tW^4 = 0$. Hence

$$a = 8f^2 \sigma^3 + T\sigma^2 \tau + 72f^3 \sigma\tau^2 + fT\tau^3,$$
$$b = -fH\sigma^4 + 18f^2 H\sigma^2 \tau^2 + HT\sigma\tau^3 + 27f^3 H\tau^4.$$

To compute c, note that (46) gives

$$x^5 - 10fx^3 + 45f^2 x - T \equiv \Pi (x - t_k),$$

identically in x. For $x = -\sigma/\tau$, this gives

$$\Pi(\sigma + \tau t_k) = \sigma^5 - 10f\sigma^3 \tau^2 + 45f^2 \sigma\tau^4 + T\tau^5.$$

But $\Pi W_k = -H^2$, the numerical factor being found by comparing one term of each member. Hence

$$c = -\Pi Y_k = H^2(\sigma^5 - 10f\sigma^3 \tau^2 + 45f^2 \sigma\tau^4 + T\tau^5).$$

To obtain a resolvent of the icosahedral equation (39), we pass to a quintic equation which involves f, H, T explicitly only in the combinations

$$Z = f^5/T^2, \qquad V = H^3/f^5,$$

where

(51) $$Z^{-1} + V = 1728$$

by the identity (29). To this end, write

(52) $$\sigma = \lambda f/H, \qquad \tau = \mu f^3/(HT).$$

Then

(53) $$\begin{cases} V a = 8\lambda^3 + \lambda^2 \mu + (72\lambda\mu^2 + \mu^3)Z, \\ V b = -\lambda^4 + 18\lambda^2 \mu^2 Z + \lambda\mu^3 Z + 27\mu^4 Z^2, \\ V c = \lambda^5 - 10\lambda^3 \mu^2 Z + 45\lambda\mu^4 Z^2 + \mu^5 Z^2. \end{cases}$$

When these values of a, b, c are inserted into (50), we obtain the *principal resolvent* of the icosahedral equation.

133. Identification of any principal quintic with the principal resolvent. Given a, b, c, we shall prove that equations (51) and (53) can be satisfied by choice of λ, μ, Z, V. To the third equation (53) add the product of the second by λ; we get the product of the first by $\mu^2 Z$, whence

(54) $$\lambda b + c = \mu^2 Z a.$$

Multiply the third equation (53) by λ and the second by $\mu^2 Z$ and subtract; we get

(55) $\qquad V(\lambda c - \mu^2 Zb) = (\lambda^2 - 3\mu^2 Z)^3.$

From the first two equations (53), we get

$$V(\lambda a + 8b)/\mu = \lambda^3 + 216\lambda^2 \mu Z + 9\lambda \mu^2 Z + 216\mu^3 Z^2.$$

Divide the square of this by Z and subtract the result from 27 times the square of the first equation (53), and simplify by use of (51); we get $V(\lambda^2 - 3\mu^2 Z)^3$. Employing (55), we get

$$27a^2 \; \frac{(\lambda a + 8b)^2}{\mu^2 Z} = \lambda c - \mu^2 Zb.$$

Elimination of $\mu^2 Z$ by means of (54) gives

(56) $\quad \begin{aligned}\lambda^2(a^4 + abc - b^3) - \lambda(11a^3 b - ac^2 + 2b^2 c) \\ + 64a^2 b^2 - 27a^3 c - bc^2 = 0.\end{aligned}$

After finding a root λ of this quadratic, we deduce $\mu^2 Z$ from (54), V from (55), and then Z from (51). By means of the resulting value of μ^2 and the first equation (53), we can find μ rationally.

The root λ involves the discriminant Δ of (56). Let P denote the product of the squares of the differences of the roots of $g(Y) = 0$, defined by (50). It is known[1] that P is the resultant of $g(Y)$ and its derivative, whence $P = 5^5 D$, where

$$D = 108a^5 c - 135a^4 b^2 + 90a^2 bc^2 - 320ab^3 c + 256b^5 + c^4.$$

We find that $\Delta = a^2 D$. In the theory of invariants, D (and not P) is called the discriminant of (50).

[1] Dickson's First Course in the Theory of Equations, 1922, 152.

Hence equations (53) and (51) may be solved rationally for λ, μ, Z, V in terms of a, b, c, $D^{\frac{1}{2}}$.

THEOREM 14. *A general principal quintic equation may be identified with the principal resolvent of the icosahedral equation. The only irrationality introduced in making this identification is the square root of the discriminant of the principal quintic.*

134. General quintic equation. By means of a transformation involving a square root, we may reduce the general quintic to a principal quintic (§118). We just proved that the latter may be identified with the principal resolvent in §132 in which λ, μ, Z, V are constants; its roots are

$$(57) \qquad Y_k = \frac{\lambda f W_k}{H} + \frac{\mu f^3 t_k W_k}{HT} \qquad (k = 0, \ldots, 4),$$

which are of degree zero in u and v and hence are polynomials in $z = u/v$. Hence the general quintic equation can be solved in terms of two square roots and a root z of the icosahedral equation (39).

The latter can be solved in terms of elliptic modular functions.[1] If with Klein we regard as known a root $z = \phi(Z)$ of the icosahedral equation, we may compute the roots of the general quintic.

135. Tschirnhaus transformation of the special Brioschi resolvent (47) into the principal resolvent of the icosahedral equation. For the system of coordinates used in §§123-6, the values of $u : v$ for which the Hessian of the octahedral form t vanishes correspond to points on the sphere which are on axes of rotations of order 3. Since the same is therefore true for the system of coordinates in §§128, 131-2, the Hessian W of the new form t

---

[1] Klein, Math. Annalen, 14, 1879, 157-8. Bianchi, *ibid.*, 17, 1880-1, 254-7. Klein, lkosaeder, 1884, 130-6; Theorie der Elliptischen Modulfunctionen, I, 1890, 125.

vanishes for values of z corresponding to mid points of faces of the icosahedron. Hence W is a factor of H. Another factor is $t^2 - 3f$. For, if we transpose the term T of (46), square and replace t^2 by $3f$, we get $3f(24f^2)^2 \equiv 1728f^5 = T^2$, whence $H = 0$ by (29). A comparison of the coefficients of one term of each member gives

(58) $$H \equiv W(t^2 - 3f).$$

From (57) with the subscripts k suppressed, (58), $t = w\,T/f^2$, and $Z = f^5/T^2$, we get

(59) $$Y = \frac{\lambda + \mu w}{Z^{-1}w^2 - 3}.$$

THEOREM 15. *The Tschirnhaus transformation* (59) *replaces resolvent* (47) *by the principal resolvent of* §132.

Since (47) is a quintic well adapted for solution by elliptic functions, we may so solve the principal resolvent and hence any quintic.

EXERCISE

Give a direct proof that (59) transforms (47) into the principal resolvent. Apply Sylvester's method to eliminate w, using (47) and its product by w, and the quadratic equation in w equivalent to (59) and its products by w, w^2, w^3, w^4. Equating to zero the determinant of these seven linear equations in $w^6, \ldots, w, 1$, show that we get (50), where a, b, c are given by (53) and (51).

136. The Galois group of the icosahedral equation. Let F_{60} denote the group of the transformations

$$z' = L_i(z) \qquad\qquad (i = 1, \ldots, 60)$$

which are the linear fractional forms of the homogeneous transformations of the icosahedral group. In view of the generators

(23) and (25) of the latter, the $L_i(z)$ involve the single irrationality ϵ. Let $I(z)$ denote the left member of the icosahedral equation (39). Its coefficients belong to the field $F = R(Z, \epsilon)$ and are invariant under the group F_{60}. Hence if z_1 is any root of $I(z) = 0$, all its roots are given by $z_i = L_i(z_1)$ for $i = 1, \ldots, 60$. Thus any factor of $I(z)$ whose coefficients are in F vanishes for all 60 roots, whence $I(z)$ is irreducible in F.

Since F_{60} is a group, $L_i[L_j(z)] \equiv L_k(z)$. Hence the replacement of z_1 by z_j gives rise to a substitution S_j on the 60 roots $z_i = L_i(z_1)$.

We shall prove that the set of these 60 substitutions S_j has the characteristic properties A and B of §89 of the Galois group of $I(z) = 0$ for F. Let $f(z_1, \ldots, z_{60})$ be any rational function of the z_i with coefficients in F. When each z_i is replaced by $L_i(z_1)$, let f become $\phi(z_1)$, whose coefficients belong to F.

First, let f be unaltered in value by each of the substitutions S_j. Then $\phi(z_1), \ldots, \phi(z_{60})$ are all equal. Their sum, which is equal to $60f$, is a symmetric function, with coefficients in F, of the roots of $I(z) = 0$. Hence f is equal to a quantity in F.

Second, let f be equal to a quantity q in F. Then $\phi(z) = q$ is an equation with coefficients in F which is satisfied by one root z_1 of the equation $I(z) = 0$, irreducible in F, and hence (§80) by all its roots z_j. In other words, the $\phi(z_j)$ are all equal, so that f is unaltered in value by all the substitutions S_j.

THEOREM 16. *The Galois group G of the icosahedral equation (39) for the field $R(Z, \epsilon)$ is simply isomorphic to the group F_{60} of the linear fractional icosahedral transformations.*

Hence, by Theorem 12, G is simply isomorphic to the alternating group on five letters. The icosahedral equation is therefore not solvable by radicals (§111).

THEOREM 17. *The solution of either the icosahedral equation or its resolvent quintic equation (47) is equivalent to the solution of the other.*

For, if we have found a root z of (39), we may compute at once the rational functions $w_k = t_k f^2/T$ of z and hence have the roots of (47). Conversely, if we have found the roots w_k of (47), we may compute z and hence all the roots of (39) as rational functions of the w_k, Z, ϵ. For by Theorem 12, the identity is the only transformation of F_{60} which leaves unaltered each t_k and hence each w_k, whence the identity is the only substitution of G which leaves each w_k unaltered. Hence the adjunction of all the w_k to the field F reduces G to the identity group, so that each root of (39) is in the enlarged field.

Exercises

1. The solution of the principal resolvent of the icosahedral equation is equivalent to the solution of the latter. Then by Theorem 17, the roots of the principal resolvent are rational functions of Z and the roots of resolvent (47), and conversely. This would have enabled us to predict the existence of a Tschirnhaus transformation which replaces (47) by the principal resolvent (§135).

2. The icosahedral equation is its own Galois resolvent.

3. For a field containing Z_1 and $(-3)^{\frac{1}{2}}$, the tetrahedral equation (32) is its own Galois resolvent, and its Galois group is simply isomorphic to F_{12} and hence to the alternating group on four letters. Thus (32) is solvable by radicals.

137. Further results. The solution of the general quintic may be reduced to the form problem of a group of 60 linear transformations on three variables.[1] The general sextic has been reduced to the form problem of Valentiner's group of 360 ternary linear transformations which is simply isomorphic to the alternating group on six letters.[2] The general equation of degree 7 reduces to the form problem of a linear group of order $\frac{1}{2}$ 7! on 4 variables.[3]

[1] Klein, Ikosaeder, 1884, 211–60.
[2] Klein, Math. Annalen, 61, 1905, 50–76 (reprinted from Jour. für Math., 129, 1905, 151–74); Math. Abhandl., II, 1922, 481–502.
[3] Klein, Math. Annalen, 28, 1887, 499–532; Math. Abhandl., II, 1922, 439–72.

The symmetric group on n letters can be represented as a linear homogeneous group on variables $x_1, \ldots, x_n$ subject to the relation $\sum x_i = 0$ (cf. §140). But[4] for $n > 7$, neither the symmetric nor the alternating group on n letters is simply isomorphic to a linear group on $n - 2$ or fewer variables. In other words the general equation of degree $n \geq 8$ can be reduced to a form problem of order $n - 1$, but not lower. For a clear summary of these and related topics, see Wiman's article[5] in *Encyklopädie Math. Wiss.*, I, i, 1904, pp. 522-54.

[4] Wiman, Math. Annalen, 52, 1899, 243-70.
[5] Later literature: E. H. Moore, Math. Annalen, 51, 1898-9, 417-44. Coble, *ibid.*, 70, 1911, 337-50; Trans. Amer. Math. Soc., 9, 1908, 396-424; 12, 1911, 311-25. Dickson, *ibid.*, 9, 1908, 121-48; 12, 1911, 75-97. Speiser, Math. Annalen, 77, 1916, 546-62. Schur, Berlin. Berichte, 1908, 664-78; Jour. für Math., 127, 1904, 20; 132, 1907, 85; 139, 1911, 155-64 (cf. de Séguier, Jour. de Math., (6), 6, 1910, 387-436; 7, 1911, 113-21). Weber, Algebra, ed. 2, 1899, II, 228-301, 373-89, 470-550.

CHAPTER XIV

REPRESENTATIONS OF A FINITE GROUP AS A LINEAR GROUP; GROUP CHARACTERS

138. Introduction. As a sequel to the preceding chapter, we shall now prove some remarkable theorems on the representations of a given group as a linear group, and also give an introduction to the theory of group characters. The latter theory is an effective tool for finite groups and has led to theorems not proved otherwise, such as the fact that every group of order $p^a q^b$ is solvable if p and q are primes, and that every transitive group of prime degree is either metacyclic or doubly transitive.

Starting with a simple example due to Dedekind, Frobenius developed the theory in a series of complicated memoirs.[1] We shall follow the simpler exposition given by Schur[2] partly because it is unusually attractive and partly since alternative introductions[3] to the subject are already available in English.[4]

139. Reducible linear groups. Let a finite number g of n-rowed non-singular matrices R form a group such that there exists a non-singular n-rowed matrix P for which

$$P^{-1}RP = M_R = \begin{pmatrix} A_R & 0 \\ C_R & D_R \end{pmatrix}$$

---

[1] Berlin. Berichte, 1896–1903. The author gave an elementary exposition of Frobenius's theory in Annals of Math., 4, 1902, 25–49; also an extension to groups of transformations modulo p, Trans. Amer. Math. Soc., 8, 1907, 389–98; Bull. Amer. Math. Soc., 13, 1906–7, 477–88.

[2] Berlin. Berichte, 1905, 406–32. Hypercomplex numbers were employed in the treatment by Molien, Sitzungsber. Naturf. Gesell. Dorpat, 11, 1896, 259–88.

[3] Blichfeldt, Finite Collineation Groups, University of Chicago Press, 1917, 116–38. Miller, Blichfeldt, and Dickson, Finite Groups, 257–78. Burnside, Theory of Groups, ed. 2, 1911, 243–371, 464–84, 499.

[4] The author added §146 and the examples.

for all g matrices R. Here A_R and D_R are square matrices with a and d rows respectively, C_R has d rows and a columns, while all elements of matrix 0 are zero. Then the group is called *reducible*.

THEOREM 1. *The matrices of a reducible group can be transformed simultaneously into*[1]

$$N_R = \begin{pmatrix} A_R & 0 \\ 0 & D_R \end{pmatrix}.$$

With R and S, the given group contains RS. Thus

$$M_{RS} = P^{-1}RSP = P^{-1}RP \cdot P^{-1}SP = M_R M_S,$$

whence

(1) $$A_{RS} = A_R A_S, \qquad D_{RS} = D_R D_S,$$

(2) $$C_{RS} = C_R A_S + D_R C_S.$$

Multiply the members of the last equation on the right by $A_{S^{-1}}$ and sum for the g values of S. Since $A_S A_{S^{-1}} = A_{I_a}$ is the a-rowed identity matrix, we get

$$\sum_S C_{RS} A_{S^{-1}} = gC_R + D_R \cdot gF, \qquad F = \frac{1}{g} \sum_S C_S A_{S^{-1}}.$$

In the first sum replace S by $R^{-1}S$. Using (1), we get

$$\sum_S C_S A_{S^{-1}} A_R = gFA_R.$$

Suppressing the common factor g, we get

[1] Maschke, Math. Annalen, 52, 1899, 363.

(3) $$FA_R = C_R + D_R F.$$

Consider the matrix of determinant unity

$$Q = \begin{pmatrix} I_a & 0 \\ F & I_d \end{pmatrix}.$$

By (3), $M_R Q = Q N_R$, where N_R is given in the theorem. Hence PQ is non-singular and transforms the given group into the group of matrices N_R.

140. Representations of a group as a linear group, group matrix. Consider a group of substitutions on n letters. Let any two of its substitutions be

$$R = \begin{pmatrix} \xi_1 & \xi_2 & \cdots & \xi_n \\ \eta_1 & \eta_2 & \cdots & \eta_n \end{pmatrix}, \quad S = \begin{pmatrix} \eta_1 & \eta_2 & \cdots & \eta_n \\ \zeta_1 & \zeta_2 & \cdots & \zeta_n \end{pmatrix}.$$

To R make correspond the linear transformation

$$r: \quad \xi_1 = \eta_1, \ldots, \xi_n = \eta_n.$$

To S corresponds s: $\eta_1 = \zeta_1, \ldots, \eta_n = \zeta_n$. Hence to RS corresponds rs: $\xi_1 = \zeta_1, \ldots, \xi_n = \zeta_n$, if we define the product rs as in §22.

THEOREM 2. *Any substitution group on n letters is simply isomorphic to a linear group on n variables.*

A substitution group may be simply or multiply isomorphic to various linear groups.

Let $s_1, \ldots, s_g$ be the substitutions of a group G (or elements of an abstract group G), whose identity is s_1. For $i = 1, \ldots, g$, let M_{s_i} be an f-rowed square matrix such that

(4) $$M_{s_i} M_{s_j} = M_{s_i s_j} \qquad (i, j = 1, \ldots, g),$$

and such[1] that not every M_{s_i} is singular. By (4) for $j = 1$, we have $M_{s_i}(M_{s_1} - I) = 0$, whence $M_{s_1} = I$, the f-rowed identity matrix. Since $M_{s_i} M_{s_i^{-1}} = M_{s_1} = I$, every M_{s_i} is non-singular.

Then the set of linear transformations having the matrices $M_{s_1}, \ldots, M_{s_g}$ is called a *representation* of G as a linear group. When the M's are not all distinct, the linear group contains duplicate transformations and is multiply isomorphic to G.

To enable us to treat all the M's simultaneously, we introduce g independent variables $x_{s_i} (i = 1, \ldots, g)$ and call

(5) $$X = \sum_{i=1}^{g} M_{s_i} x_{s_i}$$

the *group matrix* corresponding to the representation of G as the linear group $\{M_{s_1}, \ldots, M_{s_g}\}$.

For example, let $s_1 = I$, $s_2 = (12)(34)$, $s_3 = (13)(24)$, $s_4 = (14)(23)$. A group matrix is

$$X = \begin{vmatrix} x_1 & x_2 & x_3 & x_4 \\ x_2 & x_1 & x_4 & x_3 \\ x_3 & x_4 & x_1 & x_2 \\ x_4 & x_3 & x_2 & x_1 \end{vmatrix}.$$

Let T denote the transformation, of matrix X, which expresses $\xi_1, \ldots, \xi_4$ linearly in terms of $\eta_1, \ldots, \eta_4$. Introduce the new variables

$$\zeta_1 = \xi_1 + \xi_2 + \xi_3 + \xi_4, \qquad \zeta_2 = \xi_1 + \xi_2 - \xi_3 - \xi_4,$$
$$\zeta_3 = \xi_1 - \xi_2 + \xi_3 - \xi_4, \qquad \zeta_4 = \xi_1 - \xi_2 - \xi_3 + \xi_4,$$
$$\omega_1 = \eta_1 + \eta_2 + \eta_3 + \eta_4, \qquad \omega_2 = \eta_1 + \eta_2 - \eta_3 - \eta_4,$$
$$\omega_3 = \eta_1 - \eta_2 + \eta_3 - \eta_4, \qquad \omega_4 = \eta_1 - \eta_2 - \eta_3 + \eta_4.$$

---

[1] This assumption, not made by Schur, simplifies some proofs.

Then T becomes

$$\zeta_1 = (x_1 + x_2 + x_3 + x_4)\omega_1,$$
$$\zeta_2 = (x_1 + x_2 - x_3 - x_4)\omega_2,$$
$$\zeta_3 = (x_1 - x_2 + x_3 - x_4)\omega_3,$$
$$\zeta_4 = (x_1 - x_2 - x_3 + x_4)\omega_4.$$

141. Irreducible group matrices. If X is a group matrix and P is a constant matrix whose determinant is not zero, then $P^{-1}XP$ is a group matrix called *equivalent* to X.

LEMMA. *If X and Y are two irreducible group matrices of orders f and h for which $XP = PY$, where P is a constant matrix with f rows and h columns, then either $P = 0$, or else $f = h$ and the determinant of P is not zero, so that X and Y are equivalent.*

Suppose that $P \neq 0$. Then P is of rank $r > 0$. Write $s = f - r$, $t = h - r$. By §30, there exist non-singular matrices A and B with f and h rows, respectively, such that

$$Q = APB = \begin{pmatrix} I_r & Z_{rt} \\ Z_{sr} & Z_{st} \end{pmatrix},$$

where I_r is the r-rowed identity matrix, while Z_{rt} is the zero matrix with r rows and t columns. Write

$$AXA^{-1} = X_1 = \begin{pmatrix} X_{rr} & X_{rt} \\ X_{sr} & X_{st} \end{pmatrix},$$

$$B^{-1}YB = Y_1 = \begin{pmatrix} Y_{rr} & Y_{rt} \\ Y_{sr} & Y_{st} \end{pmatrix},$$

where X_{rt} has r rows and t columns, etc. We have

$$X_1 Q = AXA^{-1}Q = AXPB = APYB = APBY_1 = QY_1.$$

Hence

$$X_1 Q = \begin{pmatrix} X_{rr} & Z_{rt} \\ X_{sr} & Z_{st} \end{pmatrix} = QY_1 = \begin{pmatrix} Y_{rr} & Y_{rt} \\ Z_{sr} & Z_{st} \end{pmatrix},$$

whence $X_{sr} = 0$, $Y_{rt} = 0$. If either $s > 0$ or $t > 0$, X or Y would be reducible, contrary to hypothesis. Hence $s = t = 0$, $r = f = h$, so that $Q = I_f$ and $P = A^{-1} Q B^{-1}$ is a square matrix whose determinant is not zero.

COROLLARY. *Every constant matrix M commutative with an irreducible group matrix X of order f is of the form cI_f.*

For, if c is a root of $|M - z I_f| = 0$, $P = M - cI_f$ is a constant matrix whose determinant is zero such that $XP = PX$, whence $P = 0$.

If R is any one of the elements $s_1, \ldots, s_g$ of a group G, the corresponding f-rowed matrix A_R (previously denoted by M_R) of a representation of G as a linear group will be designated by $(a_{ij\,R})$. In a second representation of G, let R correspond to the h-rowed matrix $B_R = (b_{pq\,R})$. These notations are employed in the proof of the following

THEOREM 3. *Consider any irreducible group matrix*

$$X = (x_{ij}) \qquad (i, j = 1, \ldots, f)$$

of order f. The coefficients of

$$x_{ij} = \sum a_{ij\,R} x_R \qquad (R = s_1, \ldots, s_g)$$

satisfy the relations

(6) $$\sum_{R=s_1}^{s_g} a_{ij\,R^{-1}} a_{kl\,R} = \frac{g}{f} \partial_{il} \partial_{jk} \qquad (i, j, k, l = 1, \ldots, f),$$

where $\partial_{ii} = 1$, $\partial_{ij} = 0 \, (i \neq j)$. If

$$Y = (y_{pq}) \qquad (p, q = 1, \ldots, h)$$

is a second irreducible group matrix of order h belonging to the same group G, which is not equivalent to X, and if

$$y_{pq} = \sum b_{pqR}\, x_R \qquad (R = s_1, \ldots, s_g),$$

then

(7) $$\sum_{R=s_1}^{s_g} a_{ijR^{-1}}\, b_{pqR} = 0 \qquad \binom{i,\ j\ =\ 1,\ \ldots,\ f}{p,\ q\ =\ 1,\ \ldots,\ h}.$$

For any two elements R and S of G, we have

$$A_R A_S = A_{RS}, \qquad B_R B_S = B_{RS}.$$

Let $U = (u_{ij})$ be an f-rowed square matrix whose elements u_{ij} are arbitrary constants, and write

(8) $$V = \sum_{R=s_1}^{s_g} A_{R^{-1}}\, U A_R.$$

Then

$$A_{S^{-1}} V A_S = \sum_R A_{S^{-1}R^{-1}}\, U A_{RS}.$$

The second member is equal to V since RS ranges over the elements of the group G when R ranges over them. Since $A_{S^{-1}} A_S = A_I = I_f$, we have $V A_S = A_S V$. Hence V is commutative with every A_S and hence with X. By the Corollary, $V = v I_f$. By (8), v is a linear homogeneous function of the u's:

$$v = \sum_{j,\,k} c_{jk}\, u_{jk}.$$

Hence by (8),

$$\sum_R \sum_{j,\,k} a_{ijR^{-1}}\, u_{jk}\, a_{klR} = \partial_{il}\, v,$$

for every i, l, and for all values of u_{jk}. Hence

(9) $$\sum_R a_{ijR^{-1}} a_{klR} = \partial_{il} c_{jk} \qquad (i, j, k, l = 1, \ldots, f).$$

Take $l = i$ and sum for $i = 1, \ldots, f$; we get

$$\sum_R \sum_i a_{kiR} a_{ijR^{-1}} = fc_{jk}.$$

But $A_R A_{R^{-1}} = A_I = I_f$ implies

$$\sum_i a_{kiR} a_{ijR^{-1}} = \partial_{kj}.$$

Summing this for $R = s_1, \ldots, s_g$, we obtain the left member of the preceding equation. Hence $fc_{jk} = g\partial_{kj}$. Inserting into (9) the resulting value of c_{jk}, we get (6).

Let $W = (w_{ip})$ be a matrix with f rows and h columns whose elements are arbitrary constants, and write

$$Z = \sum_R A_{R^{-1}} W B_R.$$

As before, $A_{S^{-1}} Z B_S = Z$, whence $ZY = XZ$. Since X and Y are not equivalent, the Lemma shows that $Z = 0$. Hence

$$\sum_R \sum_{j,p} a_{ijR^{-1}} w_{jp} b_{pqR} = 0 \qquad (i = 1, \ldots, f; q = 1, \ldots, h).$$

Since the w_{jp} are arbitrary, we have (7).

THEOREM 4. *If* $X = (x_{ij})$, $Y = (y_{pq})$, ... *are any irreducible group matrices of orders* $f, h, \ldots$ *belonging to the same group* $G = \{s_1, \ldots, s_g\}$, *such that no two of the matrices are equivalent, then the* $f^2 + h^2 + \cdots$ *homogeneous linear functions* $x_{ij}, y_{pq}, \ldots$ *of the variables* x_{s_i} *are linearly independent.*

For, if the $c, d, \ldots$ are constants such that

$$\sum_{i,j} c_{ij} x_{ij} + \sum_{p,q} d_{pq} y_{pq} + \cdots = 0,$$

then in the notations of Theorem 3,

$$\sum_{i,j} c_{ij} a_{ijR} + \sum_{p,q} d_{pq} b_{pqR} + \cdots = 0.$$

Multiply by $a_{klR^{-1}}$ and sum for $R = s_1, \ldots, s_g$. By (6) and (7), we get

$$\sum_{i,j} c_{ij}\, \partial_{il}\, \partial_{jk} = 0.$$

This sum evidently reduces to c_{lk}. A similar proof gives $d_{pq} = 0$.

THEOREM 5. *The determinant of an irreducible group matrix is an irreducible function of the variables x_{s_i}. Two irreducible group matrices are equivalent if and only if their determinants are identical.*

For, if $X = (x_{ij})$ is an irreducible group matrix of order f, the x_{ij} are f^2 linearly independent functions of the x_R (Theorem 4). Hence if the u_{ij} denote f^2 independent variables, the f^2 equations $x_{ij} = u_{ij}$ determine f^2 of the x_R as homogeneous linear functions of the u_{ij} and the remaining x_R (§32). If the determinant $|x_{ij}|$ were a product of two polynomials in the x_R, then after assigning the value zero to each of those $g - f^2$ remaining x_R, we obtain a decomposition of $d \equiv |u_{ij}|$ into two polynomials v and w in the u_{ij}. Since d is of degree 1 in each variable u, we may assume that v is of degree 0 and w of degree 1 in u_{11}. No term of the expansion of the determinant d contains the product of u_{11} by an element u_{r1} of the first column. Hence v is of degree 0 in u_{r1}, whence w is of degree 1 in it. Since $u_{rc} u_{r1}$ is not a term of $v w = d$, v is of degree 0 in every u_{rc}. Hence *a determinant whose f^2 elements are independent variables is irreducible*. This completes the proof of the first part of Theorem 5.

If $X = (x_{ij})$ and $Y = (y_{pq})$ are irreducible group matrices which are not equivalent, Theorem 4 shows that we can choose the x_R so that the x_{ij} and the y_{pq} take any assigned values. Hence the determinants of X and Y are not identical.

142. Reducible group matrices. By repeated applications of Theorem 1, we see that a reducible group matrix X is equivalent to a group matrix Y in

$$(10) \quad Y = \begin{bmatrix} Y_1 & 0 & \cdots & 0 & 0 \\ 0 & Y_2 & \cdots & 0 & 0 \\ \cdots & \cdots & \cdots & \cdots & \cdots \\ 0 & 0 & \cdots & 0 & Y_m \end{bmatrix}, \quad Z = \begin{bmatrix} Z_1 & 0 & \cdots & 0 & 0 \\ 0 & Z_2 & \cdots & 0 & 0 \\ \cdots & \cdots & \cdots & \cdots & \cdots \\ 0 & 0 & \cdots & 0 & Z_n \end{bmatrix},$$

where each Y_i is an irreducible group matrix.

THEOREM 6. *Every group matrix is equivalent to one of type Y in* (10).

THEOREM[1] 7. *If a group matrix X is equivalent to both Y and Z in* (10), *where also each Z_i is irreducible, then $m = n$ and $Y_1, \ldots, Y_m$ are equivalent in some order to $Z_1, \ldots, Z_m$.*

For, let d_i and D_i be the determinants of Y_i and Z_i, respectively. By Theorem 5, d_i and D_i are irreducible. Since

$$d_1 d_2 \cdots d_m \equiv D_1 D_2 \cdots D_n,$$

and since a polynomial is decomposable into irreducible polynomials in a single way, apart from the arrangement of the factors, we have $m = n$ and see that $d_1, \ldots, d_m$ are the products of $D_1, \ldots, D_m$ in some order by constants. But for $x_I = 1$, $x_R = 0\,(R \neq I)$, the determinant of a group matrix reduces to unity. Hence the constant factors are all unity. By the last part of Theorem 5, matrices $Y_1, \ldots, Y_m$ are equivalent in some order to $Z_1, \ldots, Z_m$.

[1] For a more general such theorem, see Loewy, Trans. Amer. Math. Soc., 4, 1903, 44.

In view of Theorem 7, we shall call the group matrices $Y_1, \ldots, Y_m$ the *irreducible components* of the group matrix X. If two of them are equivalent they may be taken to be equal. Then if Y_i occurs exactly r_i times, we call r_i the *index* of Y_i, and the set of r's the indices of Y or X.

THEOREM 8. *If $X = (x_{ij})$ is a group matrix with the indices $r_1, \ldots, r_l$ and if $f_1, \ldots, f_l$ are the orders of its irreducible components, then exactly $f_1^2 + \cdots + f_l^2$ of the functions x_{ij} of the variables x_R are linearly independent. Also $r_1^2 + \cdots + r_l^2$ is equal to the number v of linearly independent constant matrices P which are commutative with X. Finally, l is equal to the number w of linearly independent constant matrices which are commutative with X and with every P.*

For, let $Y_1, \ldots, Y_m$ be the irreducible components of X, so that X is equivalent to Y in (10). The first part of our theorem follows from Theorem 4. To prove the last two parts, we may assume that the Y_i are chosen so that the first r_1 are equal, the following r_2 are equal, etc. Then let Y_i be of order t_i.

The numbers v and w remain unchanged if we replace X by Y. Let P be a constant matrix commutative with Y. We may write

$$P = \begin{bmatrix} P_{11} & P_{12} & \cdots & P_{1m} \\ \cdots\cdots\cdots\cdots\cdots\cdots \\ P_{m1} & P_{m2} & \cdots & P_{mm} \end{bmatrix},$$

where the matrix P_{ij} has t_i rows and t_j columns. Then $YP = PY$ if and only if

$$Y_i P_{ij} = P_{ij} Y_j \qquad (i, j = 1, \ldots, m).$$

If Y_i and Y_j are not equal and hence not equivalent, then $P_{ij} = 0$ by the Lemma. If they are equal, the corollary in §141 shows that $P_{ij} = c F_i$, where $F_i = I_{t_i}$ is the identity matrix of order $t_i = t_j$. Hence

$$P = \begin{pmatrix} c_{11}F_1 \cdots c_{1r_1}F_1 & 0 \cdots \cdots 0 \cdots \\ \cdots\cdots\cdots\cdots\cdots\cdots\cdots\cdots\cdots\cdots \\ c_{r_11}F_1 \cdots c_{r_1r_1}F_1 & 0 \cdots \cdots \quad 0 \\ 0 \cdots \quad 0 & d_{11}F_{r_1+1} \cdots d_{1r_2}F_{r_1+1} \\ \cdots\cdots\cdots\cdots\cdots\cdots\cdots\cdots\cdots\cdots \\ 0 \cdots\cdots 0 & d_{r_21}F_{r_1+1} \cdots d_{r_2r_2}F_{r_1+1} \\ \cdots\cdots\cdots\cdots\cdots\cdots\cdots\cdots\cdots\cdots \end{pmatrix}$$

Conversely, its origin shows that every such matrix P is commutative with X. The number of its arbitrary parameters c_{ij}, $d_{kn}, \ldots$ is $v = r_1^2 + r_2^2 + \cdots + r_l^2$.

Finally, let Q be a constant matrix commutative with X and with every P. Since Q is commutative with X, it may be derived from P by replacing the $c, d, \ldots$ by $C, D, \ldots$. Since Q shall be commutative with every P, matrix (C_{ij}) must be commutative with every matrix (c_{ij}), and (D_{kn}) with every (d_{kn}), etc. Hence $(C_{ij}), (D_{kn}), \ldots$ are products of identity matrices of orders $r_1, r_2, \ldots$ by constants. These are also sufficient conditions that Q be commutative with X and with every P. The number w of arbitrary parameters in Q is evidently l.

143. Regular group matrix. With any group $G = \{s_1, \ldots, s_g\}$ is associated a g-rowed square matrix X whose element in the ith row and jth column is x with the subscript $s_i s_j^{-1}$. Hence $X = \sum M_s x_s$, summed for all the elements s of G, where M_s denotes the matrix whose element σ_{ij} in the ith row and jth column is 1 or 0 according as $s_i s_j^{-1}$ is or is not equal to s. Similarly, M_t is the matrix whose element τ_{jk} in the jth row and kth column is 1 or 0 according as $s_j s_k^{-1}$ is or is not equal to t. In the product $M_s M_t$, the element in the ith row and kth column is $\pi_{ik} = \sum_j \sigma_{ij}\tau_{jk}$. Unless the product $\sigma\tau$ is zero, it is 1 and

$$s_i s_j^{-1} = s, \qquad s_j s_k^{-1} = t,$$

whence $s_i s_k^{-1} = st$. Hence if the latter equation does not hold, then $\pi_{ik} = 0$. If it holds, the single term of π_{ik} having a value

$\neq 0$ is that for which j is determined by $s_j = ts_k$, and then $\pi_{ik} = 1$. Hence $M_s M_t = M_{st}$.

If s_1 is the identity of G, evidently M_{s_1} is the g-rowed identity matrix. This completes the proof of

THEOREM 9. *X is a group matrix corresponding to G.*

It is called the *regular* group matrix. When G is a regular substitution group (§90), the method given at the beginning of §140 to represent G as a linear group leads to our present matrices M_{s_j}.

THEOREM 10. *Arrange the irreducible components X_i of the regular group matrix X so that no two of $X_1, \ldots, X_c$ are equivalent, while every further component is equivalent to one of these c components. If X_i is of order f_i and index e_i (so that X_i is one of exactly e_i equivalent components of X), then $f_i = e_i$ for every i, and c is the number of classes of conjugate elements of the group G of order g.*

Counting the number of rows in two ways, we have

$$e_1 f_1 + \cdots + e_c f_c = g.$$

Since each element of X is one of the g variables x, the first part of Theorem 8 gives

$$f_1^2 + \cdots + f_c^2 = g.$$

Let Y be a matrix independent of the x's. Let $[s_i, s_j]$ denote the element in the ith row and jth column of Y. By considering the element in the ith row and kth column of XY and that of YX, we see that X and Y are commutative if and only if

$$\sum_j x_{s_i s_j^{-1}} [s_j, s_k] \equiv \sum_t [s_i, s_t] x_{s_t s_k^{-1}} \qquad (i, k = 1, \ldots, g),$$

identically in the x's. The two indicated x's have equal subscripts if $s_t = s_i s_j^{-1} s_k$. By their coefficients,

$$[s_j, s_k] = [s_i, s_i s_j^{-1} s_k] \qquad (i, k, j = 1, \ldots, g).$$

Define s_n by $s_i = s_n s_j$. Then

$$[s_j, s_k] = [s_n s_j, s_n s_k] \qquad (j, k, n = 1, \ldots, g).$$

If s_1 is the identity of G, write y_{s_i} for $[s_i, s_1]$. Then

$$y_{s_j^{-1} s_i} = [s_j^{-1} s_i, s_1] = [s_i, s_j], \qquad Y = (y_{s_j^{-1} s_i}).$$

Since this matrix Y involves g independent parameters y, the second part of Theorem 8 gives

$$e_1^2 + \cdots + e_c^2 = g.$$

Hence

$$(e_1 - f_1)^2 + \cdots + (e_c - f_c)^2 = g - 2g + g = 0,$$

so that $e_i = f_i$ for every i.

A matrix Z independent of the x's which is commutative with X is of type Y. Since Y differs from X only in the order of multiplication of subscripts, a matrix commutative with every Y is of type X. Hence Z is commutative with X and with every Y if and only if its element in the ith row and jth column is simultaneously of the forms

$$z_{s_j^{-1} s_i} = z_{s_i s_j^{-1}}. \qquad (i, j = 1, \ldots, g).$$

Define s by $s_i = s_j s$. Thus $z_s = z_t$ if and only if $t = s_j s s_j^{-1}$ and hence if s and t are conjugates in G. By the final part of Theorem 8, the number c of independent parameters in Z is the number of classes of conjugate elements of G.

If possible, let there be an irreducible group matrix H of order h corresponding to G which is not equivalent to one of the c irreducible components $X_1, \ldots, X_c$ of the regular group matrix. The first part of Theorem 8 shows that in the group matrix $(x_{i,j})$ having the irreducible components $X_1, \ldots, X_c, H$,

$$f_1^2 + \cdots + f_c^2 + h^2 = g + h^2$$

of the functions x_{ij} of $x_{s_1}, \ldots, x_{s_g}$ are linearly independent, whereas evidently not more than g functions of them are linearly independent. This proves

THEOREM 11. *Corresponding to a group G, the total number of irreducible group matrices no two of which are equivalent is equal to the number c of classes of conjugate elements of G.*

144. Group characters. Consider a representation of $G = \{s_1 = I, s_2, \ldots, s_g\}$ as a group of linear transformations with the non-singular matrices M_{s_i}. Denote the *trace* (sum of the diagonal elements) of the latter by $\chi(s_i)$. The set of g numbers $\chi(s_1), \ldots, \chi(s_g)$ is called a *character* of G corresponding to this representation or to the group matrix $X = \sum M_{s_i} x_{s_i}$ belonging to it.

For the 4-rowed group matrix X in the example in §140, M_{s_1} is the identity matrix, whence $\chi(s_1) = 4$; while M_{s_2} is obtained from X by taking $x_2 = 1$, $x_i = 0 (i \neq 2)$, whence $\chi(s_2) = 0$. We saw that $\zeta_2 = z \omega_2$, where $z = x_1 + x_2 - x_3 - x_4$. If we employ this new group matrix (z) having a single element z, we obtain the new character $\chi(s_1) = \chi(s_2) = 1$, $\chi(s_3) = \chi(s_4) = -1$.

Since equivalent matrices have the same trace, to equivalent representations or equivalent group matrices correspond the same character. By Theorem 11, there exist irreducible group matrices $X_1, \ldots, X_c$, no two of which are equivalent, such that every irreducible group matrix is equivalent to one of them. The corresponding (simple) characters are denoted by $\chi_1(s_i), \ldots, \chi_c(s_i)$.

If among the irreducible components of the group matrix X, X_j occurs exactly r_j times $(r_j \geqq 0)$, evidently

(11) $\qquad \chi(s_i) = r_1 \chi_1(s_i) + \cdots + r_c \chi_c(s_i).$

Let u and v be two distinct numbers of the set $1, \ldots c$, and write

$$X_u = \sum_{R=s_1}^{s_g} (a_{ijR}) x_R, \quad X_v = \sum_R (b_{pqR}) x_R \quad \begin{pmatrix} i, j = 1, \ldots, f_u \\ p, q = 1, \ldots, f_v \end{pmatrix}.$$

Then

$$\chi_u(R) = \sum_i a_{iiR}, \qquad \chi_v(R) = \sum_p b_{ppR}.$$

In (6) replace i by m and k by n, multiply the resulting equation by $a_{imS} a_{knT}$ and sum for $m, n = 1, \ldots, f$. We get

$$\sum_R \sum_{m=1}^{f} a_{imS} a_{mjR^{-1}} \cdot \sum_{n=1}^{f} a_{knT} a_{nlR} = \frac{g}{f} \sum_{m,n} \partial_{ml} \partial_{jn} a_{imS} a_{knT}.$$

The summand on the right is zero unless $m = l$, $n = j$. Since $A_S A_{R^{-1}} = A_{SR^{-1}}$, we may perform the summations on the left and obtain

(12) $$\sum_R a_{ijSR^{-1}} a_{klTR} = \frac{g}{f} a_{ilS} a_{kjT}.$$

From (7) we obtain in the same manner

(13) $$\sum_R a_{ijSR^{-1}} b_{pqTR} = 0.$$

In (12) take $j = i$, $l = k$, $T = I$; we get

$$\sum_R a_{iiSR^{-1}} a_{kkR} = \frac{g}{f} a_{ikS} \partial_{ki}.$$

Sum for i and k, and mark f with the subscript u. Thus

(14) $$\sum_R \chi_u(SR^{-1}) \chi_u(R) = \frac{g}{f_u} \chi_u(S).$$

Since $\chi_u(I) = f_u$, the case $S = I$ gives

(15) $$\sum_R \chi_u(R^{-1}) \chi_u(R) = g.$$

Similarly, in (13) take $j = i$, $q = p$, $T = I$, and sum for i, p.
We get

(16) $$\sum_R \chi_u(SR^{-1})\chi_v(R) = 0 \qquad (u \neq v).$$

We return to the general group matrix X to which corresponds the character $\chi(s_i)$ for $i = 1, \ldots, g$. Multiply (11) by $\chi_u(s_i^{-1})$ and sum for $i = 1, \ldots, g$. We get

$$\sum_{i=1}^{g} \chi(s_i)\chi_u(s_i^{-1}) = \sum_{v=1}^{c} \sum_{i=1}^{g} r_v \, \chi_v(s_i)\chi_u(s_i^{-1}).$$

For $v \neq u$, the inner sum on the right is zero by (16) with $S = I$. For $v = u$, it is $r_u \, g$ by (15). Hence r_u is determined by the values of $\chi(s_i)$ and the fixed numbers $\chi_u(s_i^{-1})$. But r_j is the number of times X_j occurs as an irreducible component of X. This proves

THEOREM 12. *Two representations of G as linear groups are equivalent if and only if their corresponding characters are the same.*

145. Applications to group matrices. Let G be a commutative group, and $X = \sum M_{s_i} x_{s_i}$ be a corresponding irreducible group matrix of order f. Since M is commutative with X, it is of the form $c_i I_f$ by the corollary in §141. If s_i is of order e_i, $M^{e_i} = I_f$, whence c_i is an e_ith root of unity. If $f > 1$, $X = \sum (c_i x_{s_i}) I_f$ would be a reducible matrix.

THEOREM 13. *An irreducible group matrix corresponding to a commutative group is of order unity.*

See the example at the end of §140.

It follows from Theorem 6 that every representation of a commutative group of order g is equivalent to one whose matrices have gth roots of unity in the diagonal and zeros everywhere outside the diagonal. This property therefore holds for the special case of a cyclic group and proves

268   REPRESENTATION AS LINEAR GROUP   [Ch. XIV

THEOREM 14. *The trace of any linear transformation of finite order is a sum of roots of unity.*

THEOREM 15. *The order of every irreducible group matrix of G is a divisor of the order g of G.*

Write $E_s = 1$ or 0 according as $s = I$ or $s \neq I$. Then

$$\sum_R \chi(R) E_{SR^{-1}} = \chi(S),$$

so that (14) may be written in the form

$$\sum_R \chi(R) \left\{ \frac{g}{f} E_{SR^{-1}} - \chi(SR^{-1}) \right\} = 0.$$

Since $\chi(I) = f \neq 0$, the determinant

$$\left| \frac{g}{f} E_{SR^{-1}} - \chi(SR^{-1}) \right| = 0 \qquad (S, R = s_1, \ldots, s_g).$$

We saw that the trace χ of SR^{-1} is a sum of roots of unity and hence is an integral algebraic number. Thus g/f is a root of an equation of degree g whose leading coefficient is unity and remaining coefficients are integral algebraic numbers. Hence it is an integral algebraic number. Being at the same time a rational number, it is a rational integer.

146. The alternating group G on five letters. By Ex. 5, §111, G has exactly five sets of conjugate substitutions. Let the five representations of G as irreducible linear groups have the orders 1, a, b, c, d. They are divisors of 60 and the sum of their squares is 60. Their maximum is evidently 6. If $d = 6$, $a^2 + b^2 + c^2 = 23$, which is impossible in integers. Let $d = 5$. Then $a^2 + b^2 + c^2 = 34$. If $c = 5$, $a^2 + b^2 = 9$ and a or b is zero. If $c = 4$, $a = b$

= 3. Not all of a, b, c are ≤ 3. Finally, if a, b, c, d are all ≤ 4, not all are 4, while $1 + 4^2 + 4^2 + 4^2 + 3^2 = 58 < 60$.

THEOREM 16. *Every representation of the alternating group on five letters as an irreducible linear group is equivalent to one on 1, 3, 4, or 5 variables, there being two types on 3 variables.*

147. Computation of group characters. In (12) take $k = j$, $l = i$, and sum for i and j; we get

(17) $$f\sum_R \chi(SR^{-1}TR) = g\chi(S)\chi(T).$$

Since the traces of $A_R A_S$ and $A_S A_R$ are equal,

(18) $$\chi(RS) = \chi(SR).$$

Consider the symmetric group on 3 letters and write

$$E = \text{identity}, A = (132), B = (123), C = (12), D = (13),$$
$$F = (23).$$

The three sets of conjugates are E; A, B; C, D, F. By (18),

$$\chi(A) = \chi(B), \quad \chi(C) = \chi(D) = \chi(F).$$

By (17) for $T = E$, $\chi(E) = f$. By (17) for $S = T = A$; $S = T = C$; $S = A, T = C$ in turn, and by (15), we get

$$3f^2 + 3f\chi(A) = 6\chi^2(A), \quad 2f^2 + 4f\chi(A) = 6\chi^2(C),$$
$$6f\chi(C) = 6\chi(A)\chi(C), \quad f^2 + 2\chi^2(A) + 3\chi^2(C) = 6.$$

If $\chi(C) \neq 0$, the third relation gives $\chi(A) = f$, and the other relations give $\chi^2(C) = f^2 = 1$. Since the order f of the irreducible group matrix is positive, we have

$$f = 1, \quad \chi(E) = \chi(A) = 1, \quad \chi(C) = \pm 1.$$

If $\chi(C) = 0$, the conditions reduce to $\chi(A) = -\tfrac{1}{2}f$, $f^2 = 4$, whence

$$f = 2, \quad \chi(E) = 2, \quad \chi(A) = -1, \quad \chi(C) = 0.$$

SUBJECT INDEX

References are to pages

Abel's theorem, 178
Adjunction, 151, 174, 180, 190, 195, 197
Alternant, 27
Alternating function, 147
Archimedes' problem, 207

Bilinear forms, 51–3, 60 (see Hermitian)
 alternate, 81, 125–6
 canonical, 59
 canonical pair, 115–9
 equivalent, 57, 65
 equivalent pairs, 112–5, 122–5, 134
 matrix of, 52, 53
 pencil of, 119, 120
 symmetric, 64–6, 71, 73, 124
Binary form, 3 (see canonical)
Binomial equation, 156, 199

Canonical form of binary cubic, 9
 of equations, 210–19
 of form even order, 38
 of form odd order, 36
 of quartic, 20–2
Canonizant, 37
Chain of functions, 90
Characteristic determinant, 47, 93–7
 equation, 47, 48, 76
 matrix, 47
Commutator, 27
Conic pairs, 132–3
Conjugate numbers, 66
 quaternions, 46
Conjugates to function, 172, 183–5
Covariant, 5–14, 16–20, 24–38
 annihilators of, 24–33
 as invariants, 34
 differential operators producing, 32–3
 functionally complete system of, 38
 fundamental system of, 17–20, 32–5
 homogeneous, 10, 11
 index of, 4, 5

Covariant, isobaric, 13
 leader of, 13–7, 29, 30
 of binary cubic, 19
 of binary quadratic, 18
 of binary quartic, 20
 of transformations modulo p, 38, 82
 resolvent of quintic, 203
 tables of, 23
 weights of coefficients of, 12, 13
 with given leader, 16, 29, 30, 37
Cubic, binary, 7–9
 equation, 9, 135–8, 180, 194, 214
 form, 3 (see discriminant)
Cycle, 137
Cyclotomic equation, 156–8, 191–4, 208

Dependent, 6, 54
Determinants, 3, 6, 37, 55, 127–8
 irreducible, 259
 minor of product of two, 49
 rank of, 49
 with linear elements, 134
Differential equations, 134, 177
Discriminant of binary cubic, 7–9, 18
 of cubic equation, 138–9
 of quadratic form, 4, 18
 of quartic, 143–4
 of quintic, 245
Domain of rationality, 54
Duplication of cube, 204, 207

Elementary divisors, 109, 110, 113–5, 117–9, 125–6, 132
Elliptic functions, 217, 220, 246–7
Equal functions, 159
Equation (see elliptic, general, group of, icosahedral)
 normal forms of, 212–8
 solved by auxiliary equations, 180, 189, 193, 209
 with cyclic group is solvable, 190
 with root constructible geometrically, 205

271

SUBJECT INDEX

References are to pages

Equation, with roots powers of those of another, 211
Euler's theorems, 7, 28, 192

Field, 54, 150
  modular, 107
Form, 2
  problems, 236–9, 249–50
Function with $n!$ values, 160, 163
Functional determinant, 6

Galois resolvent, 162–7, 174, 249
Galois's criterion for solvability, 198
  theorem on adjunction, 197
Gauss's lemma, 154
General equation, 249, 250
  not solvable by radicals, 178, 200–2
Geometrical constructions, 204–9
Greatest common divisor, 151–2
Group (see group matrix, subgroup)
  abstract, 144–5
  alternating, 148–9, 176, 182, 186, 189, 200–3, 226, 240, 248–50, 268–9
  characters, 265–70
  composite, 187
  cyclic, 144, 149, 185
  equation with prescribed, 176
  factors of composition of, 188–9, 202
  function belonging to, 146, 170
  Γ on conjugates, 181–5, 187
  generated, 144, 146
  icosahedral, 232–5, 247–8
  identity, 144
  insolvable, 189
  intransitive, 168
  isomorphic, 182
  leaving function invariant, 145
  linear, 249–51, 253
  monodromie, 176–7
  octahedral, 229–31, 236–9
  of binomial equation, 199
  of cyclotomic equation, 192
  of De Moivre's quintic, 177
  of equation, 159, 164–7, 174–7, 195
  of equations in geometry, 177
  of general equation, 175–6
  of icosahedral equation, 247
  of regular solids, 220
  of resolvent equation, 181–3

Group of substitutions, 144, 253
  order of, 144, 149, 168
  quotient, 187–8, 196
  reducible, 252
  regular, 168, 185
  regular cyclic, 185, 189, 190, 192–4, 197, 199, 208–9
  representation as linear, 251–69
  series of composition of, 188–9, 202
  simple, 187–8, 200–2
  solvable, 189, 193, 198, 251
  symmetric, 144, 175–6, 186, 189, 202, 231, 250
  tetrahedral, 223–9, 236–8
  transform of, 184
  transitive, 168, 182, 251
Group matrix, 254–68
  equivalent, 255
  index of, 261
  irreducible components, 261
  regular, 263

Hermitian bilinear forms, 66–8, 73, 122–5
  alternate, 81, 125
Hermitian forms, 67–76, 82–8, 102, 134
  equivalent, 68, 73
  equivalent pairs of, 112, 122–5
  index of, 71, 73
  negative, 73
  number of positive terms, 88
  pair of, 74, 134
  pair with any invariant factors, 126–30
  positive, 73
  reduction of, 68–73, 82–8
  signature, 77
  under unitary transformation, 76, 124
Hessian, 3–5, 7–10, 12, 19, 20, 22, 23, 230, 235, 242, 246
Hilbert's theorem, 30, 231
Homogeneous, 2, 28

Icosahedral equation, 239, 246
  Brioschi's resolvent of, 242, 246–8
  group of, 247–8
  principal resolvent of, 242–7, 249
Icosahedron, 232
Identity element of group, 144

SUBJECT INDEX

References are to pages

Identity substitution, 139, 141
  transformation, 2
Invariant factors, 97, 103–6, 109–10, 113–8, 124–32
Invariants, 5, 17
  absolute, 21
  annihilators of, 26, 33
  coefficients of canonical form are, 38
  differential operator producing, 32, 33
  fundamental system of, 32–3, 231
  index of, 4, 5
  of icosahedral group, 235–6, 238
  of octahedral group, 229–31, 236–9
  of quadratic, 4
  of quartic, 19, 20, 22, 195
  of system of covariants, 7, 8
  of tetrahedral group, 227–9, 236–8
  of two quadratic forms, 7, 119
  weight of, 16
Irreducible, 153–8, 168, 170, 172, 176
Isobaric, 13, 28

Jacobian, 6, 7, 19, 20, 231, 235
Jordan's theorem on adjunction, 195

Leader of chain, 90
Linear equations, homogeneous, 60–2
  non-homogeneous, 62, 63
Linear forms, 3, 39–41
  independent, 55, 56, 59, 60
Linear fractional transformation, 220, 223, 225–6, 234
Linear transformation, 1, 16, 41, 226
  associative law, 2
  classic canonical form, 105–10
  cogredient, 64
  conjugate, 66
  determinant of, 1, 42
  generators of, 16, 26
  identity, 2, 43
  in new variables, 98, 103–4
  inverse, 2, 43, 99
  leaving quadratic surface invariant, 110–1
  non-singular, 42
  orthogonal, 76, 77, 98–102, 111, 124–5
  product, 1, 42

Linear transformation, rational canonical form, 89–97, 103
  similar, 104
  singular, 42
  unitary, 76, 77, 124
  with any invariant factors, 103
Linearly independent, 54–6, 61, 62

Matrices, 39–53
  adjoint, 44
  associative law of, 45
  characteristic, 47
  congruent, 65, 133
  conjugate, 66
  conjunctive, 67, 133
  determinants of, 49
  distributive law of, 45
  elements matrices, 47
  elements of, 39
  elements polynomials in λ, 134
  equal, 40
  equivalent, 57, 123
  equivalent pairs of, 112–5, 134
  greatest common divisor of determinants of, 96, 97, 102–4, 106
  group of, 251–69
  Hermitian, 66, 77–81
  identity, 43
  inverse, 44, 46
  nth roots of, 120–1
  product of, 41
  rank of, 49–51, 53, 57, 59, 60, 63, 77–80
  regular, 81
  scalar, 43, 46
  similar, 104
  skew-symmetric, 80
  sum of, 45
  symmetric, 64, 65, 77–81
  trace of, 265, 268
  transpose of, 53
  zero, 43

Non-singular case, 115
Normal form of equations, 210–9

Octahedron, 229

p-ic, 3
Primitive roots of unity, 157–8, 191–2
Principal equation, 211–2, 219, 244–6
  minor, 79–81

SUBJECT INDEX

References are to pages

Properties A and B of group of equation, 165–6

q-ary, 3
Quadratic forms, 3, 7, 65–76, 102, 134
  canonical form of, 70–2, 82–8
  canonical pair of, 74, 126–33
  definite, 73
  equivalent, 65, 70, 71, 134
  equivalent pairs, 112, 122–5
  index of, 71, 72, 88
  Kronecker's reduction of, 82–7
  law of inertia of, 72
  negative, 73
  number of positive terms, 87
  pair with any invariant factors, 126–32
  positive, 73
  reduction of, 68–72
  regular, 87
  under orthogonal transformation, 76, 77, 124–5
Quadratic function a square, 142–3
Quantity, 150
Quartic equation, 23, 142–4, 194–5, 214
  discriminant of, 143–4
  reciprocal, 159, 177
  resolvent cubic for, 143
Quartic form, 3
Quaternary form, 3
Quaternion, 46, 47, 110–1
Quintic equation, general, 246, 249
  of Bring and Jerrard, 214
  of Brioschi, 217 (see icosahedral)
  solvable case, 203

Rational function of roots, 163, 165–7, 170–4
  operations, 179
Reducible, 153
  linear groups, 251
Regular polygons, 204, 207–9
Relatively prime, 152
Removal of three terms from an equation, 212–3
Resolvent equations, 180–3, 203 (see Galois, icosahedral)

Roots, integral, 155
  of equations defined, 152–3
  of unity (see cyclotomic)
Rotations, 98–101, 110, 222–3, 225–36, 239–41

Seminvariant, 14–20, 27, 29, 37, 38
Solvability by radicals, 179, 190, 193–4, 198
Special points, 228, 231, 235–6
Stereographic projection, 220
Subgroup, 144, 174
  complete set of conjugate, 185, 203
  index of, 149
  invariant, 176, 184–7, 196–7
  maximal invariant, 187–8
  order of, 148
  proper, 187
  rational function belonging to, 171
  self-conjugate, 185
Substitution, 137, 139 (see group)
  associative law, 140–1
  even, 147–8, 186
  identity, 141
  invariant, 187
  inverse, 142
  negative, 147
  odd, 147–8
  order, 141, 149
  positive, 147, 186
  power of, 141
  product, 140
  transform of, 184, 186
  two-rowed notation, 139
Sums of squares, product of two, 133
Symbolic notation, 38
Syzygy, 19, 20, 35, 229, 231, 236

Ternary form, 3
Tetrahedral equation, 249
Tetrahedron, 223
  diametral, 223
Transposition, 146–7
Trisection of angle, 204, 206–7
Tschirnhaus transformation, 210–9, 247, 249

Unary form, 3

Weight, 12, 13, 16

AUTHOR INDEX

References are to pages

Abel, 178
Archimedes, 207

Baker, 177
Bauer, 176
Berwick, 203
Bianchi, 177, 246
Blichfeldt, 251
Bliss, 80
Bôcher, 134
Börger, 177
Breuer, 176
Bring, 213
Brioschi, 203, 217-8, 242
Bromwich, 133
Brown, 188
Bucca, 196
Burkhardt, 220
Burnside, 251

Cayley, 35, 39, 102, 203
Clebsch, 38
Coble, 250
Coolidge, 153

Dedekind, 251
De Moivre, 156, 177
Descartes, 205
De Séguier, 250
Dickson, 38, 54, 80, 82, 89, 105, 111-2, 125, 133-4, 177, 203, 207, 209, 245, 250-1

Elliott, 38
Euler, 7, 28, 178, 192

Faà di Bruno, 23
Ferrari, 142
Frobenius, 112, 126, 134, 177, 251
Furtwängler, 176

Galois, 159, 162, 179, 197-8
Garver, 219
Gauss, 154, 209
Glenn, 38

Gordan, 7, 35, 38, 218
Grace, 38

Hamilton, 178, 213
Hermite, 64, 177
Hesse, 3, 7
Hilbert, 30, 176, 231
Hilton, 133
Hölder, 196
Hurwitz, 111, 133, 177

Jacobi, 6, 72, 203
Jerrard, 213
Jordan, 105, 188, 195

Kiepert, 217
Klein, 220, 246, 249
Kneser, 177, 196
König, 177
Kowalewski, 89
Kronecker, 82, 126, 134
Krull, 89

Lagrange, 173, 191
Landsberg, 196
Lattès, 89
Loewy, 102, 177, 196, 260
Logsdon, 112

Maillet, 176
Maschke, 252
Mathews, 102, 203
McClintock, 203
Meyer, 38
Miller, 188
Molien, 251
Moore, 250
Muth, 102, 125-6, 133

Nöther, E., 176
Nöther, M., 7

Pascal, 102
Perrin, 203
Picard, 177

275

AUTHOR INDEX

References are to pages

Radon, 133
Rasor, 220
Ritt, 177
Ruffini, 178
Schlesinger, 205
Schur, 176, 250-1, 254
Scott, 102
Seidelmann, 176
Serret, 178
Speiser, 250
Stickelberger, 126
Sylvester, 72

Tschirnhaus, 210, 212

Valentiner, 249
Vessiot, 177

Wäisälä, 176
Wantzel, 178
Weber, 153, 176, 218, 250
Wedderburn, 80
Weierstrass, 89, 112, 131, 134
Weisner, 177
Weitzenböck, 38
Wilczynski, 38
Wiman, 250

Young, 38

DOVER PHOENIX EDITIONS

A series of hardcover reprints of major works in mathematics, science and engineering.
All editions are 5⅝ × 8½ unless otherwise noted.

Mathematics

Theory of Approximation, N. I Achieser. Unabridged republication of the 1956 edition. 320pp. 49543-4
The Origins of the Infinitesimal Calculus, Margaret E. Baron. Unabridged republication of the 1969 edition. 320pp. 49544-2
A Treatise on the Calculus of Finite Differences, George Boole. Unabridged republication of the 2nd and last revised edition. 352pp. 49523-X
Space and Time, Emile Borel. Unabridged republication of the 1926 edition. 15 figures. 256pp. 49545-0
An Elementary Treatise on Fourier's Series, William Elwood Byerly. Unabridged republication of the 1893 edition. 304pp. 49546-9
Substance and Function & Einstein's Theory of Relativity, Ernst Cassirer. Unabridged republication of the 1923 double volume. 480pp. 49547-7
A History of Geometrical Methods, Julian Lowell Coolidge. Unabridged republication of the 1940 first edition. 13 figures. 480pp. 49524-8
Linear Groups with an Exposition of Galois Field Theory, Leonard Eugene Dickson. Unabridged republication of the 1901 edition. 336pp. 49548-5
Continuous Groups of Transformations, Luther Pfahler Eisenhart. Unabridged republication of the 1933 first edition. 320pp. 49525-6
Transcendental and Algebraic Numbers, A. O. Gelfond. Unabridged republication of the 1960 edition. 208pp. 49526-4
Lectures on Cauchy's Problem in Linear Partial Differential Equations, Jacques Hadamard. Unabridged reprint of the 1923 edition. 320pp. 49549-3
The Theory of Branching Processes, Theodore E. Harris. Unabridged, corrected republication of the 1963 edition. xiv+230pp. 49508-6
The Continuum, Edward V. Huntington. Unabridged republication of the 1917 edition. 4 figures. 96pp. 49550-7
Lectures on Ordinary Differential Equations, Witold Hurewicz. Unabridged republication of the 1958 edition. xvii+122pp. 49510-8
Mathematical Methods and Theory in Games, Programming, and Economics: Two Volumes Bound as One, Samuel Karlin. Unabridged republication of the 1959 edition. 848pp. 49527-2
Famous Problems of Elementary Geometry, Felix Klein. Unabridged reprint of the 1930 second edition, revised and enlarged. 112pp. 49551-5
Lectures on the Icosahedron, Felix Klein. Unabridged republication of the 2nd revised edition, 1913. 304pp. 49528-0
On Riemann's Theory of Algebraic Functions, Felix Klein. Unabridged republication of the 1893 edition. 43 figures. 96pp. 49552-3
A Treatise on the Theory of Determinants, Thomas Muir. Unabridged republication of the revised 1933 edition. 784pp. 49553-1
A Survey of Minimal Surfaces, Robert Osserman. Corrected and enlarged republication of the work first published in 1969. 224pp. 49514-0
The Variational Theory of Geodesics, M. M. Postnikov. Unabridged republication of the 1967 edition. 208pp. 49529-9

DOVER PHOENIX EDITIONS

An Introduction to the Approximation of Functions, Theodore J. Rivlin. Unabridged republication of the 1969 edition. 160pp. 49554-X

An Essay on the Foundations of Geometry, Bertrand Russell. Unabridged republication of the 1897 edition. 224pp. 49555-8

Elements of Number Theory, I. M. Vinogradov. Unabridged republication of the first edition, 1954. 240pp. 49530-2

Asymptotic Expansions for Ordinary Differential Equations, Wolfgang Wasow. Unabridged republication of the 1976 corrected, slightly enlarged reprint of the original 1965 edition. 384pp. 49518-3

Physics

Semiconductor Statistics, J. S. Blakemore. Unabridged, corrected, and slightly enlarged republication of the 1962 edition. 141 illustrations. xviii+318pp. 49502-7

Wave Propagation in Periodic Structures, L. Brillouin. Unabridged republication of the 1946 edition. 131 illustrations. 272pp. 49556-6

The Conceptual Foundations of the Statistical Approach in Mechanics, Paul and Tatiana Ehrenfest. Unabridged republication of the 1959 edition. 128pp. 49504-3

The Analytical Theory of Heat, Joseph Fourier. Unabridged republication of the 1878 edition. 20 figures. 496pp. 49531-0

States of Matter, David L. Goodstein. Unabridged republication of the 1975 edition. 154 figures. 4 tables. 512pp. 49506-X

The Principles of Mechanics, Heinrich Hertz. Unabridged republication of the 1900 edition. 320pp. 49557-4

Thermodynamics of Small Systems, Terrell L. Hill. Unabridged and corrected republication in one volume of the two-volume edition published in 1963–1964. 32 illustrations. 408pp. 6½ x 9¼. 49509-4

Theoretical Physics, A. S. Kompaneyets. Unabridged republication of the 1961 edition. 56 figures. 592pp. 49532-9

Quantum Mechanics, H. A. Kramers. Unabridged republication of the 1957 edition. 14 figures. 512pp. 49533-7

The Theory of Electrons, H. A. Lorentz. Unabridged reproduction of the 1915 edition. 9 figures. 352pp. 49558-2

The Principles of Physical Optics, Ernst Mach. Unabridged republication of the 1926 edition. 279 figures. 10 portraits. 336pp. 49559-0

The Scientific Papers of James Clerk Maxwell, James Clerk Maxwell. Unabridged republication of the 1890 edition. 197 figures. 39 tables. Total of 1,456pp.
Volume I (640pp.) 49560-4; Volume II (816pp.) 49561-2

Vectors and Tensors in Crystallography, Donald E. Sands. Unabridged and corrected republication of the 1982 edition. xviii+228pp. 49516-7

Principles of Mechanics and Dynamics, Sir William (Lord Kelvin) Thompson and Peter Guthrie Tait. Unabridged republication of the 1912 edition. 168 diagrams. Total of 1,088pp. Volume I (528pp.) 49562-0; Volume II (560pp.) 49563-9

Treatise on Irreversible and Statistical Thermophysics: An Introduction to Nonclassical Thermodynamics, Wolfgang Yourgrau, Alwyn van der Merwe, and Gough Raw. Unabridged, corrected republication of the 1966 edition. xx+268pp. 49519-1

Engineering

Principles of Aeroelasticity, Raymond L. Bisplinghoff and Holt Ashley. Unabridged, corrected republication of the original 1962 edition. xi+527pp. 49500-0

Statics of Deformable Solids, Raymond L. Bisplinghoff, James W. Mar, and Theodore H. H. Pian. Unabridged and corrected Dover republication of the edition published in 1965. 376 illustrations. xii+322pp. 6½ x 9¼. 49501-9